D0214850

INTERNATIONAL SERIES OF MONOGRAPHS IN THE SCIENCE OF THE SOLID STATE

GENERAL EDITOR: B. R. PAMPLIN

VOLUME 2

II–VI COMPOUNDS

A

II–VI COMPOUNDS

BY

BRIAN RAY

Lecturer in Electronics,
Queen's College, University of St. Andrews

PERGAMON PRESS

OXFORD · LONDON · EDINBURGH · NEW YORK
TORONTO · SYDNEY · PARIS · BRAUNSCHWEIG

PERGAMON PRESS LTD.,
Headington Hill Hall, Oxford
4 & 5 Fitzroy Square, London W.1

PERGAMON PRESS (SCOTLAND) LTD.,
2 & 3 Teviot Place, Edinburgh 1

PERGAMON PRESS INC.,
Maxwell House, Fairview Park, Elmsford, New York 10523

PERGAMON OF CANADA LTD.,
207 Queen's Quay West, Toronto 1

PERGAMON PRESS (AUST.) PTY. LTD.,
19a Boundary Street, Rushcutters Bay, N.S.W. 2011, Australia

PERGAMON PRESS S.A.R.L.,
24 rue des Écoles, Paris 5e

VIEWEG & SOHN GMBH,
Burgplatz 1, Braunschweig

First edition 1969

Library of Congress Catalog Card No. 72-93126

Printed in Great Britain by Neill & Co., Ltd. of Edinburgh.

08 006624 0

CONTENTS

PREFACE

The purpose of this text is to provide the reader with a general idea of the way in which II–VI compounds behave. The size of the book is such as to limit the detail with which individual materials can be described and the discussion to those compounds with a reasonably well understood behaviour pattern. Parts of the text have been used for honours and diploma courses in Materials Science in the Department of Electrical Engineering at the University of St. Andrews, and as such are suggestive of the background knowledge necessary for the reader. The text is of necessity complementary to more general and extensive works, particularly in the fields of luminescence and photoconductivity, where the II–VI compounds have found most application up to the present time.

The book begins with a general chapter on the type of bonding which is thought to occur and the crystal structures that are observed along with a generalized categorization of the relevant energy band structures. A chapter on the preparation of the II–VI compounds both as pure and as usefully activated materials precludes four chapters on physical properties. These four chapters are divided mainly on the basis of the research efforts and hence show a strong bias towards the optical sensitivity of the II–VI compounds. A final chapter on the application of II–VI compounds considers what has already been achieved and looks optimistically at some future possibilities. There are obvious omissions particularly in the fields of physical chemistry and magnetic resonance properties of these compounds. However, discussions on such subjects would extend the text so greatly as to defeat the book's purpose as a means of quick reference. Reference to these untreated subjects is given at appropriate points in the text.

The author is greatly indebted to the many people who have contributed either directly or indirectly to this work. In particular I am indebted to Professor J. C. Woolley of the University of Ottawa and Dr. D. W. G. Ballentyne of the Imperial College, London, who introduced me to many different aspects of the study of II–VI compounds. My thanks go to the many research workers who have freely provided me with reprints of their published work and greatly assisted the task of reference. I wish also to thank Dr. B. R. Pamplin and Dr. P. M. Spencer for their criticisms of the text and Miss S. Urquhart and Mr. D. R. Allan for the preparation of the final manuscripts. Finally I would like to thank my wife, Susan, not only for her forbearance while this text was being prepared but also for her contribution in the preparation of draft manuscripts.

BRIAN RAY

ACKNOWLEDGEMENTS

I SHOULD like to thank the publishers, editors and authors of the following journals and books for permission to reproduce line diagrams and tables.

The *Physical Review* and the *Physical Review Letters* (American Institute of Physics and the American Physical Society) ; *Journal of Applied Physics, Applied Physics Letters, Journal of the Optical Society of America* (American Institute of Physics) ; *Applied Optics* (Optical Society of America); *Philips Research Reports*; *Journal of the Physical Society of Japan*; *Journal of the Electrochemical Society*; *General Telephones* and *Electronics Journal*; *Physica Status Solidi* (Academic Press, Inc.); *Proceedings of the Seventh International Conference on the Physics of Semiconductors* (Dunod and Academic Press, Inc.); *Sylvania Technologist*; *Journal of Electronics* (Taylor and Francis Ltd.); *Proceedings of the Institute of Electrical and Electronic Engineering*; *Proceedings of the Royal Society*; *Journal of the Physics and Chemistry of Solids, Solid State Communications, Solid State Electronics, Journal of Inorganic and Nuclear Chemistry* (Pergamon Press Ltd.).

Physics and Chemistry of II–VI Compounds edited by M. Aven and J. S. Prener (North Holland Publishing Company); *Photoconductivity of Solids* by R. H. Bube (John Wiley and Sons, Inc.); *Luminescence in Inorganic Solids* edited by P. Goldberg (Academic Press, Inc.); *Progress in Semiconductors* edited by A. F. Gibson (Heywood Books Ltd.); *Photoelectronic Materials and Devices* edited by S. Larach (D. Van Nostrand Co., Inc.).

LIST OF SYMBOLS AND ABBREVIATIONS

m	meter
g	gram
s	second
e	electronic charge
A	ampere
V	volt
W	watt
eV	electron volts
Å	Ångstrom units
l	litre
torr	pressure equal to 1/760 of an atmosphere
°C	degrees centigrade
°K	degrees absolute
Ω	Ohm

E	energy
$\mathbf{E}$	electric field vector
ε	magnitude of the electric field
I	current
J	current density
V	voltage
t	time
T	temperature
β	heating rate
κ	thermal conductivity

κ_e electronic component of thermal conductivity
κ_L lattice component of thermal conductivity
P pressure
p_A partial pressure of element A
ε_S static dielectric constant
ε_∞ optical or high frequency dielectric constant
ε_O permittivity of free space
k Boltzmann's constant
h Planck's constant
$\hbar = h/2\pi$
$\mathbf{k}$ wave vector
λ wavelength
υ frequency
$\omega = 2\pi\upsilon$ angular frequency
α absorption coefficient
σ electrical conductivity
ρ electrical resistivity
R_H Hall coefficient
μ_H Hall mobility
μ_n electron mobility
μ_p hole mobility
b electron-to-hole mobility ratio
n density of electrons
p density of holes
m_o free electron mass
m^* effective mass
m_p polaron mass
τ free carrier or recombination lifetime
v thermal velocity of charge carriers
q^* ionicity
E_c energy at the bottom of the conduction band
E_v energy at the top of the valence band
$E_G = E_c - E_v$ magnitude of the forbidden energy gap
E_F Fermi energy
Δ_{so} spin-orbit splitting energy
Δ_{cr} crystalline field splitting energy

s carrier frequency escape probability factor from traps

N_C density of electron states in the conduction band

N_V density of electron states in the valence band

B magnetic flux density

A.C. alternating current

D.C. direct current

mol. molecular

K_p reaction or equilibrium constant

$\sim$ of the order of

$>$ greater than

$\gg$ very much greater than

$<$ less than

$\ll$ very much less than

CHAPTER 1

FUNDAMENTAL NATURE OF II–VI COMPOUNDS

1.1. Introduction

II–VI compounds as a collective group of materials have been and still are the subject of much intensive investigation. Along with the III–V compounds, the II–VI compounds have represented an alternative course of study to that of the elemental group IV semiconductors. The growth of semiconductor technology in the early 1950's highlighted the limitations of silicon and germanium of which perhaps the character and magnitude of the forbidden energy gap were the most disadvantageous. At first the extension in the range of energy gaps was sought in the III–V compounds, where considerable success has been achieved with InSb and GaAs in the low- and high-energy gap areas respectively. GaAs is today probably the most developed and well-understood compound in existence. Concurrently with the later developments in the III–V compounds, systematic studies were made of several of the II–VI compounds. The results of these studies have revealed much about the general nature of the II–VI compounds and the feature of chemical stability of the higher energy gap materials at room temperature offers an immediate advantage over the unstable III–V phosphides. Direct energy gaps seem to be a general character of the II–VI compounds with the exception of the semi-metals, HgTe and HgSe, and augur well for their use in devices which require high absorption or emission of radiation in the region of the energy gap. Variety in the usefulness of II–VI compounds also makes them an interesting family of compounds as instanced by the semi-metallic HgTe mentioned above, highly photoconducting CdS and CdSe and strongly luminescent ZnS. As the technology of the already established compounds grows, so it is hoped that their full potential will be realized and further efforts will be directed to the study of the more obscure II–VI compounds, such as the chalcogenides of beryllium and magnesium. It is with this background view that the text has been drawn up to highlight the important characteristics of the II–VI compounds and to contrast differences in their behaviour. The author makes no claim to the book being comprehensive although he hopes that it will transmit many concepts and much information to the reader in the short space available.

It is necessary in an introduction to define formally what is implied in the title and set appropriate limitations to the scope of the text. II–VI compounds in their broadest sense include compounds formed from elements of group II and group VI of the periodic table; such a definition encompasses the oxides, sulphides, selenides and tellurides of beryllium, magnesium, zinc, cadmium and mercury. Fortunately there exists an almost natural limit to the compounds to be discussed here in that the chalcogenides of beryllium and magnesium plus the oxides of cadmium and mercury have received relatively little attention. Thus the general discussion in the book will centre around the sulphides, selenides and tellurides of zinc, cadmium and mercury and zinc oxide, although occasional reference may be made to the materials excluded in the previous sentence. The restricted range of II–VI compounds made above take one of two crystalline structures, zinc blende and wurtzite, both of which are characterized by tetrahedral lattice sites. The sodium chloride structure is taken by some of the lesser known compounds which have been excluded from the general discourse.

1.2. Crystallographic Form

1.2.1. *Lattice Sites*

The combination of group II and VI elements gives on average four valence electrons per atom; this is a situation conducive to the formation of tetrahedral lattice sites provided there is a tendency towards sharing rather than the transfer of electrons between atoms. A tetrahedral lattice site in a compound AB is one in which each atom A is surrounded symmetrically by four nearest neighbouring B atoms. For this to occur, the B atoms must sit on the corners of a tetrahedron with the A atom situated at its geometrical centre. The A and B sites are identical as far as their tetrahedral nature is concerned. The combination of the tetrahedral sites takes two possible forms which are relevant to the compounds of interest. Figure 1.1(a) illustrates the situation when the base triangles of the interpenetrating tetrahedra are parallel and lined up normal to each other. Figure 1.1(b) shows the base triangles again parallel but rotated through 60° about the normal to each other.

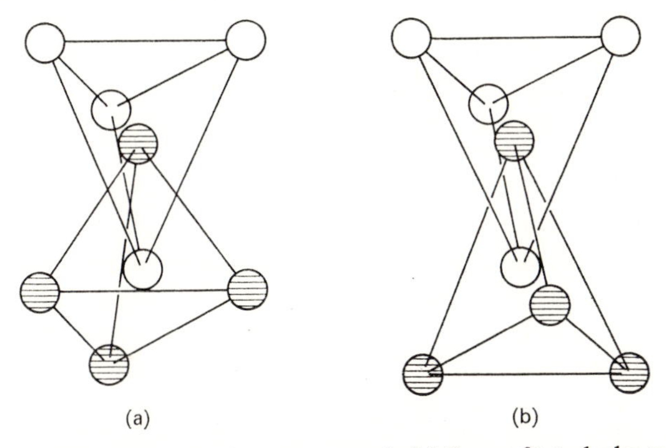

FIG. 1.1. Tetrahedral sites for a compound. (a) Bases of tetrahedra parallel and in line vertically. (b) Bases of tetrahedra parallel and 60° out of line vertically.

1.2.2. *Crystal Structures*

These two combinations of the tetrahedral lattice sites lead to the two crystal structures, wurtzite and zinc blende.

(a) *Wurtzite.* The wurtzite structure which is in the hexagonal crystal class has the combination of tetrahedral sites illustrated in Fig. 1.1(a). It consists of two interpenetrating close-packed hexagonal lattices, as illustrated in Fig. 1.2, displaced with respect to each other by a distance $\frac{3}{8}c$ along the hexagonal c-axis. The nearest neighbour distance in the wurtzite structure with ideal tetrahedral sites is $\frac{3}{8}c$ or $\sqrt{\frac{3}{8}}a$, which gives a c/a ratio of $\sqrt{8/3} = 1{\cdot}632$. BeO, ZnO, ZnS, CdS, ZnSe, CdSe and MgTe have all been observed to take the wurtzite structure.

(b) *Zinc blende.* The zinc blende structure which is in the cubic crystal class has the combination of tetrahedral sites illustrated in Fig. 1.1(b). It is derived from the diamond structure and is composed of two interpenetrating cubic close-packed lattices, illustrated in Fig. 1.3, translated with respect to each other by $\frac{1}{4}$ of the body diagonal. The nearest neighbour separation in this instance is $\frac{\sqrt{3}}{4}a$. The sulphides,

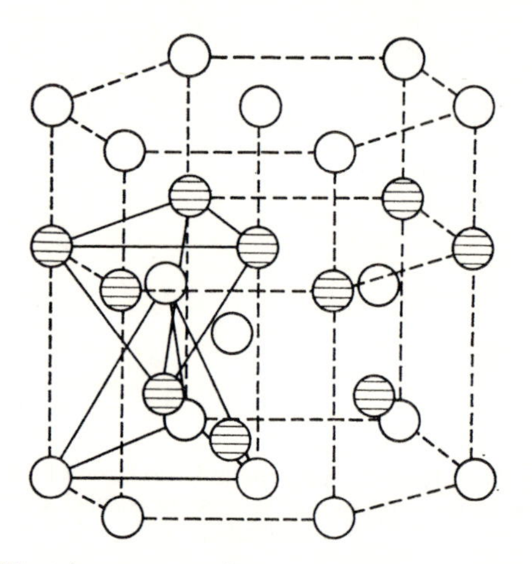

FIG. 1.2. Wurtzite structure, illustrating type (a) tetrahedral site.

selenides and tellurides of beryllium, zinc, cadmium and mercury have all been obtained with the zinc blende structure.

The two structures can be compared in terms of the [111] threefold axis of zinc blende and the [00·1] hexad axis of wurtzite. Rotation of the interpenetrating tetrahedra in zinc blende about the [111] axis converts the structure to wurtzite and the symmetry axis becomes a [00·1] axis. Comparison of the electronegativity differences of the two atoms which form a compound suggests that the wurtzite structure is more favourable to compounds with larger atomic electronegativity differences.

(c) *Sodium chloride.* This structure consists of a combination of two interpenetrating cubic close-packed lattices translated with respect to each other by half the length of the cube edge, illustrated in Fig. 1.4. Tetrahedral bonding is not present in the NaCl structure, which has only octahedral lattice sites and is generally associated with a crystal having a predominantly ionic bond. At high pressures the wurtzite and zinc blende structures of many of the II–VI compounds have been observed to revert to the sodium chloride structure, which is a more tightly packed structure. CdO, MgO, MgS and MgSe take the sodium chloride structure at room temperature.

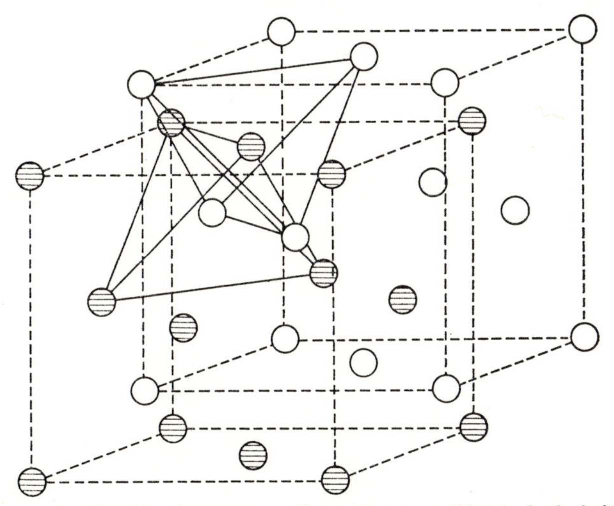

FIG. 1.3. Zinc blende structure, illustrating type (b) tetrahedral site.

1.3. Bonding Mechanisms

1.3.1. *Covalent Bond*

The tetrahedral lattice sites which characterize the zinc blende and wurtzite structures suggest that homopolar bonding is present. The homopolar bond is more simply understood in terms of a tetravalent atom where the electronic wave functions are taken to be sp^3-hybridized and can combine in four independent ways to give the highly directional character of the bond, which is directed towards the corners of a tetrahedron. The sp^3-hybridization requires that the s^2p^2 state which normally exists in the group four atoms is modified to the sp^3 state. In the formation of the crystal from the isolated atoms the energy required for hybridization and binding is lost. In II–VI compounds with an average of four valence electrons per atom available for bonding the same basic mechanism can occur; however, this is necessarily modified in the final result, since the ions which occupy the lattice sites differ in charge. For example in CdTe, Cd^{2+} and Te^{6+} ions are the atomic core ions, which compare with the C^{4+} ions in diamond a material that has a purely homopolar bond.

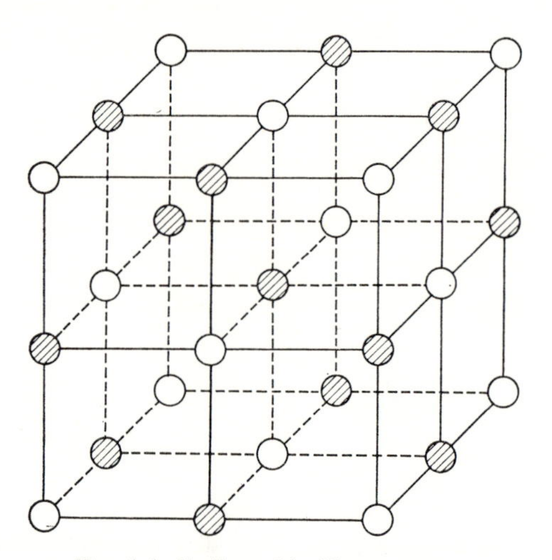

FIG. 1.4. Sodium chloride structure.

1.3.2. *Mixed Covalent–Ionic Bond*

There is the possibility that instead of the electron spin pairing which characterizes the homopolar bond there may be an exchange of electrons. This results in an ionic electrostatic attraction which characterizes the ionic bond. Take, for example, an ionic bond in CdTe, it would give the two ions Cd^{2+} and Te^{2-} occupying the lattice sites. However, it is clear from the structures taken by the II–VI compounds that the homopolar bond is present to a considerable degree although the ionic bond does coexist with it in varying degrees. The ionic character increases as the atomic weight decreases, i.e. from HgTe to ZnS. The ionic bond has the effect of strengthening the lattice, raising the melting point and increasing the energy gap.

Pauling[1] has assigned covalent radii to the atoms concerned and compared these with the values obtained experimentally from the lattice parameter. Further comparison can be made with the nearest neighbour separation which is computed from the ionic radii. It should be noted that a correction has to be made to the ionic separation since the coordination number is four in zinc blende and not the value of six for which the ionic radii are quoted. Table 1.1 lists the covalent and ionic radii of the relevant elements and the nearest neighbour separations determined theoretically and experimentally.

TABLE 1.1

Atom	At. No.	Covalent radii, Å	Ionic radii, Å		Nearest neighbour separation, Å		
					Covalent compound	Ionic compound	Lattice parameter observed experimentally
Zn	30	1·31	0·74	ZnS	2·35	2·42	2·38
Cd	48	1·48	0·97	ZnSe	2·45	2·55	2·46
Hg	80	1·48	1·10	ZnTe	2·63	2·76	2·64
S	16	1·04	1·84	CdS	2·52	2·63	2·52
Se	34	1·14	1·98	CdSe	2·62	2·76	2·62
Te	52	1·32	2·21	CdTe	2·80	2·98	2·79
Si	14	1·17	—	HgS	2·52	2·75	2·53
Ge	32	1·22	—	HgSe	2·62	2·89	2·63
Sn	50	1·40	—	HgTe	2·80	3·11	2·80
Pb	82	1·46	—				

A comparison between the experimental and theoretical values tends to suggest most of the II–VI compounds have a strongly covalent bond. This agreement between theory and experiment, however, must only be accepted with reservations, since measurements of electrical properties indicate that a certain amount of ionic character is necessary to explain the reduced carrier mobilities.

1.3.3. *Neutral Bond*

The difference in the charges at the adjacent atomic cores in the II–VI compounds, which in CdTe would be Cd^{2+} and Te^{6+}, present the opportunity to consider another viewpoint for the bonding. In this a neutral bond is taken as the centre of discussion and has been considered for the III–V compounds by Madelung.(2) If the electron bridge between Cd^{2+} and Te^{6+} has its centre of charge at the midpoint, since the valence electrons screen the ionic charges, the lattice atoms have

TABLE 1.2. BOND SCHEME AFTER MOOSER AND PEARSON(4)

Compound	Pure covalent bond	Neutral bond
$A^{IV}A^{IV}$ [Ge]	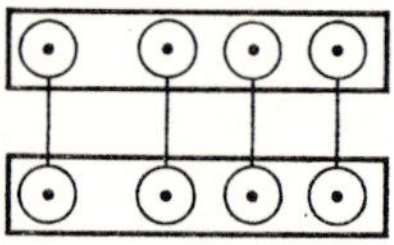	Unchanged
$A^{II}B^{VI}$ [CdTe]	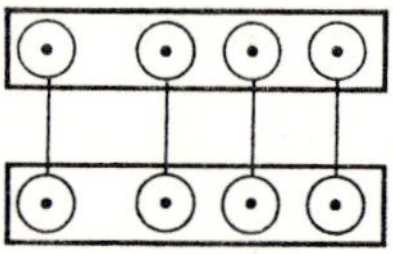	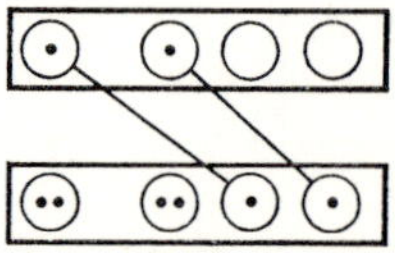

an effective charge of $\pm 2e$. The homopolar bonding is thus characterized by $Cd^{2-}Te^{2+}$ which compares with $Cd^{2+}Te^{2-}$ for a purely ionic bond. Transitions between these two states are seen to occur continuously when the bonding is considered over many molecules. It is clear that when the electron bridge moves nearer to the Te atoms a point will be reached at which the effective lattice atomic charge is zero. This gives the case of neutral bonding described by $Cd^{0}Te^{0}$.

The bonding that is discussed above is seen to be both homopolar and neutral and compares exactly with that of the group IV elements which take the diamond structure. Thus a II–VI compound fulfilling these conditions would be similar in nature to the group IV element or compound of the same average atomic number. In practice, it is possible to show that the bonding in II–VI compounds is approximately neutral. Table 1.2 compares the covalent and neutral bonds in terms of the sharing of s and p electrons.

1.3.4. *Representation of the Chemical Bond Character*

A quantitative estimation of the bond character based on work by Coulson, Redei and Stocker[(3)] will be given below in abbreviated form. Homopolar bonding of two nearest neighbours A and B is described by the combination of the sp^3 hybrid wave functions $\phi_{A,B}$ of the two atoms.

$$\psi = \phi_A + \phi_B \tag{1.1}$$

If ionic bonding occurs the equation becomes

$$\psi = \lambda\phi_A + \phi_B \tag{1.2}$$

where λ is a parameter related to ionicity of the bond.

The effective ionic charge of the two atoms is obtained using λ. Electron spends $\lambda^2/(1+\lambda^2)$ of its time on atom A and $1/(1+\lambda^2)$ upon atom B. Since there are on average 4 electrons per atom, then the charge due to these electrons on atom A is

$$-\frac{8e\lambda^2}{1+\lambda^2},$$

and thus the effective charge e_A^* on A is

$$e_A^* = 2e - \frac{8e\lambda^2}{1+\lambda^2} = \frac{2-6\lambda^2}{1+\lambda^2}.e$$

and the effective charge of atom B is $e_B^* = -e_A^*$. λ varies from 1 to 0 and hence $e_A^* = -2e$ for homopolar bonding and $e_A^* = +2e$ for ionic bonding. Neutral bonding occurs when $e_A^* = 0$, $\lambda = 0{\cdot}577$. Table 1.3 notes the effective charge on A atoms of $A^{II}B^{VI}$ compounds with the zinc blende structure as obtained by Coulson *et al.*[3] These values have been calculated by the method of linear combination of atomic orbitals (LCAO).

TABLE 1.3

Compound	e_A^*/e
ZnO	0·60
ZnS	0·47
CdS	0·49
HgS	0·46
ZnSe	0·47
CdSe	0·49
HgSe	0·46
ZnTe	0·45
CdTe	0·47
HgTe	0·44

Mooser and Pearson[4] have introduced a method of expressing quantitatively the type of bonding. The quantity used is the ionicity which refers to the percentage ionic character of the bond. The pure ionic state has an ionicity of 100% and the pure covalent state has a formal ionicity of zero. There is an additional method which is often used to define the ionicity and this gives the neutral bond state an ionicity of zero and is known as the effective ionicity. In the case of a II–VI compound the neutral bond state would have a formal ionicity of 50%. Table 1.4 compares the formal and effective ionicities for a range of diamond-like semiconductors.[5]

1.4. Energy Band Structure Related to the Brillouin Zone

1.4.1. *Diamond and Zinc Blende Structures*

The energy band structures of germanium and silicon are probably most familiar to the reader.[6] These are derived from the Brillouin zone of the diamond structure taken by germanium and silicon. The Brillouin zone of the zinc blende structure is identical to that of

TABLE 1.4. FORMAL AND EFFECTIVE IONICITY OF DIAMOND-LIKE SEMICONDUCTORS EXPRESSED AS A PERCENTAGE[4]

Compound type	Effective charge of the lattice sites							Definition
	$A^{-3}B^{+3}$	$A^{-2}B^{+2}$	$A^{-1}B^{+1}$	$A^{0}B^{0}$	$A^{+1}B^{-1}$	$A^{+2}B^{-2}$	$A^{+3}B^{-3}$	
$A^{IV}A^{IV}$ (Ge)				0				
$A^{III}B^{V}$ (GaAs)			0	25	50	75	100	Formal ionicity
				0	33	67	100	Effective ionicity
$A^{II}B^{VI}$ (ZnSe)		0	25	50	75	100		Formal ionicity
				0	50	100		Effective ionicity
$A^{I}B^{VII}$ (CuBr)	0	25	50	75	100			Formal ionicity
				0	100			Effective ionicity

diamond, which is represented in Fig. 1.5 and takes the form of a truncated octahedron. The principal symmetry points and lines are indicated in this figure and characterize translational properties in **k**-space. Γ is the zone centre, Λ is the [111] axis and L its intersection with the zone edge, Δ is the [100] axis and X its intersection with the zone edge.

The translational properties are insufficient to describe fully the symmetry in **k**-space. Rotational axes, reflection planes and inversion points are symmetry elements necessary for a complete representation. In order to study this subject fully group theory is required, a discussion of which is beyond the scope of this text; however, it is of importance

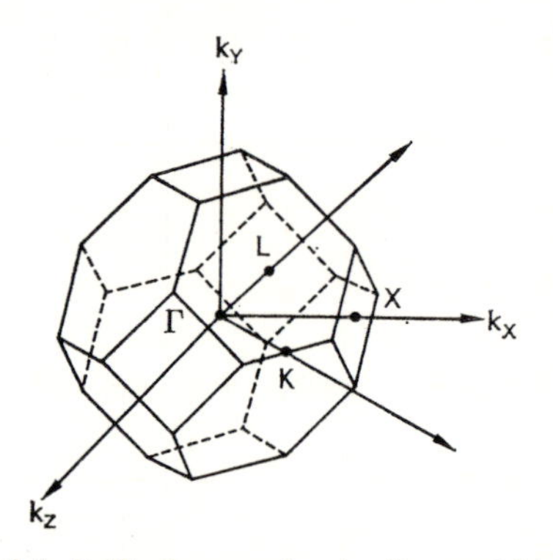

FIG. 1.5. Brillouin zone for the diamond lattice.

to note that these additional symmetry elements determine directly the curvature of the energy bands. One final matter which has to be mentioned in connection with the symmetry is the spin; it causes degeneracy of energy levels since spins of opposite signs may have the same energy. It is possible to employ the same symbols for energy levels in zinc blende crystals as in diamond crystals; this is despite the fact that the inversion centre of the diamond structure is lost in the zinc blende structure. The heteropolar potential which results may be treated simply as a perturbation of that in diamond.[7, 8]

In the diamond structure the valence band is composed of eight sub-bands. At $\mathbf{k} = 0$ there are two bands forming the upper edge

Γ_{25}, which is threefold orbitally degenerate (p^3) or sixfold spin degenerate, and the lower edge Γ_1, which is twofold spin degenerate (s). The upper edge is of further interest since the consideration of spin orbit interaction effects splits it into a fourfold spin degenerate term with Γ_8 symmetry and a twofold spin degenerate term with Γ_7 symmetry. Γ_7 is lower in energy than Γ_8 by the spin–orbit interaction energy Δ_{so}. The Γ_8 term splits into two bands away from $\mathbf{k} = 0$, which are both approximately parabolic but of different curvatures. The two curvatures result in a heavy hole V_1-band and a light hole V_2-band; the Γ_7 term forms the top of the V_3-band. When the approach is extended to the zinc blende structure if the loss of inversion centre is treated merely as a perturbation effect the same symmetry designation may be used.

The conduction band at $\mathbf{k} = 0$ has two minima, $\Gamma_{2'}$ and Γ_{15}, which are two- and sixfold spin degenerate respectively. For the zinc blende structure, if a non-perturbational approach is taken the $\Gamma_{2'}$ term becomes the Γ_1 term. $\Gamma_{2'}$ level is associated with s electrons and Γ_{15} with p^3 electrons. The other positions in $\mathbf{k}$-space where the lowest minima have been observed are the [111]-axis, Λ_1 or L_1 (s electrons) and Λ_3 or L_3 (p^2 electrons), and [100] axis, Δ_1 or X_1 (sp electrons). In the group IV elements the lowest minimum varies between Γ, L and Δ minima,[6, 10] whereas in the II–VI compounds which have been studied to any extent the lowest minimum appears to be at the Γ point.[9]

1.4.2. *Wurtzite Structure*

The wurtzite and zinc blende structures are both characterized by tetrahedral lattice sites and as a consequence their nearest neighbour environment is identical. Only slight differences exist in the next nearest neighbour situation and it is necessary to go beyond next nearest neighbours before both positional and directional differences occur. The crystal field effect which results from such differences is sufficiently small to permit general comparison of the energy bands in zinc blende and wurtzite.[11] The changes in band structure which occur in the transition from zinc blende to wurtzite can be simply discussed in terms of the two Brillouin zones. Birman[11] has shown that, in order to make a direct comparison between the two zones, a double zone scheme must be used for the wurtzite structure, as shown in Fig. 1.6. This doubling is necessitated by the fact that the number of atoms per unit cell in wurtzite is twice that in zinc blende.

The important symmetry points in the wurtzite double zone are Γ and Γ' points which are the centres of the single zones translated through half of the c-axis separation of the single zone. The midpoint of Γ–Γ', A, and the [00·1] axis, Δ, are significant in the comparison

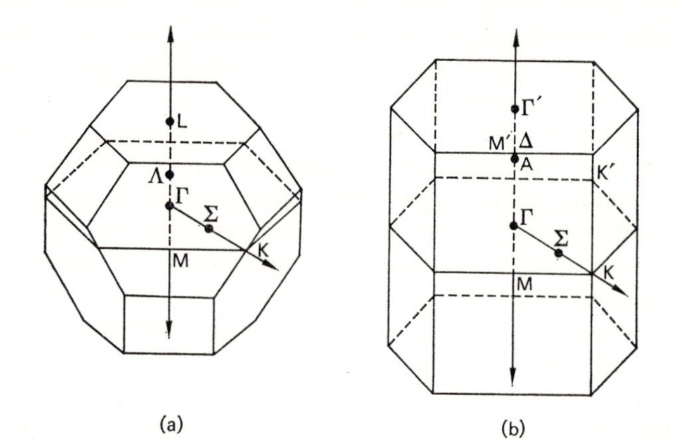

FIG. 1.6. (a) Brillouin zone for zinc blende. (b) Double Brillouin zone for wurtzite after Cardona and Harbeke.[12]

with the zinc blende structure. The following other symmetry conditions for wurtzite are used in the tabulated comparison with diamond and zinc blende given at the end of the chapter: the intersection of the [1$\bar{1}$·0] axis with the zone edge at ($\frac{1}{2}$, $\frac{1}{2}$, 0) gives the M symmetry point; the [11·0] axes are described by a Σ symmetry notation; the intersection of the [10·0] axis with the zone edge at (1, 0, 0) gives the K symmetry point.

The correspondence between zinc blende and wurtzite is obtained in the ΓL and $\Gamma A \Gamma'$ directions. Clearly the correspondence only holds for two of the eight equivalent ΓL directions of zinc blende ± [111];

it is not possible to define electronic states along the other six directions ($\pm[\bar{1}11]$, $\pm[1\bar{1}1]$, $\pm[11\bar{1}]$) of zinc blende by the perturbational approach. We have mentioned above that the number of atoms per unit cell in wurtzite is twice that in zinc blende and as a result the number of Γ states in wurtzite is twice that in zinc blende. Half of the

TABLE 1.5. COMPARABLE TRANSITIONS IN k-SPACE EXPECTED FOR DIAMOND, ZINC BLENDE AND WURTZITE STRUCTURES FROM REFLECTIVITY SPECTRA (AFTER CARDONA AND HARBEKE[12])

	Diamond	Zinc blende	Wurtzite $E \parallel C$	Wurtzite $E \perp C$
E_0	$\Gamma_{25'} \to \Gamma_{2'}$	$\Gamma_{15} \to \Gamma_1$	$\Gamma_1 \to \Gamma_1$	$\left.\begin{matrix}\Gamma_5\\ \Gamma_1\end{matrix}\right\}\Gamma_1$
E_0'	$\Gamma_{25'} \to \Gamma_{15'}$	$\Gamma_{15} \to \Gamma_{15}$	$\Gamma_5 \to \Gamma_5$ $\Gamma_1 \to \Gamma_1$	$\Gamma_5 \to \Gamma_1$ $\Gamma_1 \to \Gamma_5$
E_1 A B	 $\Lambda_3 \to \Lambda_1$	 $\Lambda_3 \to \Lambda_1$	Forbidden $M_1 \to M_1$, $K_2 \to K_2$, $\Sigma_1 \to \Sigma_1$	$\Delta_5 \to \Delta_1$ $M_2 \to M_4$, $M_1 \to M_4$, $\Sigma_1 \to \Sigma_2$
E_1'	$L_{3'} \to L_3$	$L_3 \to L_3$	$\Gamma_6 \to \Gamma_6$ and others	Unknown
E_2	$X_4 \to X_1$	$X_5 \to X_1$	Unknown	Unknown
E_2'		$X_5 \to X_3$	Unknown	Unknown
d_1	$L_{3'} \to L_{2'}$	$L_3 \to L_1$	Unknown	$\Gamma_6 \to \Gamma_3$ and others
d_2	From *d*-electrons ⟶			
F_1	—	—	Forbidden	Unknown
F_2	—	Unknown	—	—
F_3	—	—	Unknown	Unknown

Γ wurtzite states correspond directly with the Γ zinc blende states. The threefold orbitally degenerate $\Gamma_{25'}$ (Γ_{15}) energy states in zinc blende correspond to twofold degenerate Γ_5 and singly degenerate Γ_1 states in wurtzite. The Γ_5–Γ_1 split at the top of the valence band is derived from the absorption of radiation polarized with the electric vector **E** parallel and perpendicular to the *c*-axis; splitting of the order of 0·030 eV is observed and is called the crystalline field splitting which results from the anisotropic nature of the wurtzite structure. Certain Γ states in wurtzite do not correspond to states in zinc blende and this is a consequence of the use of the double zone in order to compare the two structures. In wurtzite the Γ_6 and Γ_3 states correspond to L_3 and L_1 states in zinc blende.

Table 1.5 gives a much fuller comparison of the energy transitions which occur in the three structures, diamond, zinc blende and wurtzite. It considers transitions in addition to those for the principal extrema in the conduction and valence bands. The subscripts which are used with the energy E observed from reflection spectra in diamond and zinc blende structures denote the following: 0 for transitions at the Γ point with $\mathbf{k} = 0$, 1 for transitions between Λ or L positions, i.e. on the [111] axis, 2 for transitions between points on the [100] axis. The prime notation on the E' refers to a higher energy transition than E at the same symmetry point.

The experimental determination of the band structure and the results that are obtained for the different II–VI compounds will be discussed in Chapter 3 on fundamental optical properties. A full review article on the band structures in the II–VI compounds is given by Segall.[13]

References

1. PAULING, L., *The Nature of the Chemical Bond*, 3rd edition, Cornell U.P., New York, 1960.
2. MADELUNG, O., *Physics of III–V Compounds*, 1st edition, John Wiley, New York, 1964.
3. COULSON, C.A., REDII, L. B., and STOCKER, D., *Proc. Roy. Soc.* **270,** 357–72 (1962).
4. MOOSER, E., and PEARSON, W. B., *J. Electronics* **1,** 629–45 (1956).
5. FOLBERTH, O.G., *Compound Semiconductors* **1,** 23–33, Reinhold, New York, 1962.
6. HERMAN, F., *Rev. Mod. Phys.* **30,** 102–21 (1958).
7. PHILIPS, J. C., *Solid State Physics* **18,** 76–101, Academic Press, New York, 1966.
8. NUSSBAUM, A., *Solid State Physics* **18,** 261–72, Academic Press, New York, 1966.
9. COHEN, M. L., and BERGSTRESSER, T. K., *Phys. Rev.* **141,** 789–96 (1966).
10. GROVES, S., and PAUL, W., *Phys. Rev. Letters* **11,** 194–6 (1963).
11. BIRMAN, J. L., *Phys. Rev.* **115,** 1490–2 (1959).
12. CARDONA, M., and HARBEKE, G., *Phys. Rev.* **137A,** 1467–76 (1965).
13. SEGALL, B., *Physics and Chemistry of II–VI Compounds*, pp. 3–72. (M. Aven and J. S. Prener eds.) North Holland, Amsterdam, 1967.

CHAPTER 2

PREPARATION AND SINGLE-CRYSTAL GROWTH

THE essence of technological advance lies in the ability of manufacturers to produce materials of extremely high quality, both from the point of view of foreign impurities and natural defects. High purities in the II–VI compounds are now within reach. One of the features, which has inhibited the development of the II–VI compounds, is their generally high melting points. Cadmium sulphide and zinc sulphide are typical of such a characteristic in this class of materials in that they have found application in the radiation sensitive field since the turn of the century but it has only been in the form of polycrystalline powders. However, in the past decade single crystals of cadmium sulphide have become more readily available and have led to rapid advances in the compound's development; piezoelectric effect and laser emission have both been observed in high-quality cadmium sulphide crystals. Although great strides have been made with the sulphides because of their obvious application potential, it is the tellurides that have received a really thorough systematic investigation; no doubt this is because they have lower melting points than other members of the family. In particular the work of de Nobel, Lorenz and their respective coworkers on CdTe and Harman and his coworkers on HgTe exemplify these studies. In this chapter the pattern of the above developments will be discussed and presented successively in terms of the purification of the elements, phase equilibria in the binary systems, the preparation and single-crystal growth of the compounds and the preparation of thin films.

2.1. Purification of the Elements

The elements of the II–VI compounds which will be considered in this text in terms of purification techniques are zinc, cadmium, mercury, oxygen, sulphur, selenium and tellurium. All of these elements have relatively low melting points which simplifies their purifying treatment. Table 2.1 gives the melting and boiling points of these elements.

TABLE 2.1. MELTING AND BOILING POINTS OF THE GROUP II AND GROUP VI ELEMENTS

Element	M.P. (°C)	B.P. (°C)
Zn	420	908
Cd	321	765
Hg	− 38·5	365
O	− 220	− 183
S	119	445
Se	217	685
Te	450	987

Oxygen, since it is gaseous at room temperature, is the only element not subjected to standard purification techniques. It is purified by passage through various absorbent solutions and finally a drying agent.

Vacuum distillation is a general technique which may be applied to the remainder of the elements. In the technique volatile impurities are transported and condensed out by a cold trap while less volatile impurities are left at the heat source. The distillation process for mercury requires gentle movement from the high- to low-temperature region. It is important that bumping of the mercury does not occur, for if it does, the splashing of mercury results and the dissolved particles are carried to the cold region. With zinc and cadmium the distillation requires a furnace with a temperature gradient and a sectional pyrex tube so that the purified material automatically isolates itself.

If further purification is required for the elements that exist in the solid form the method of zone refining can be applied to all except selenium.[1] In the method the impurities that form a solid solution with the element are removed. The technique originated by Pfann[2] uses a zone which is moved horizontally along the length of a solid ingot, Fig. 2.1. The ingot of the element is either contained in a tube under vacuum or in an inert gas atmosphere. The molten zone sweeps some impurities along with it to the far end of the ingot; the remaining impurities are left behind in the solid and after several passes of the molten zone begin to accumulate at the first end to freeze. It is the distribution coefficient of the impurity in the host element that determines the direction in which the impurities move. Repeated sweeps of the zone in the same direction produce a very pure central

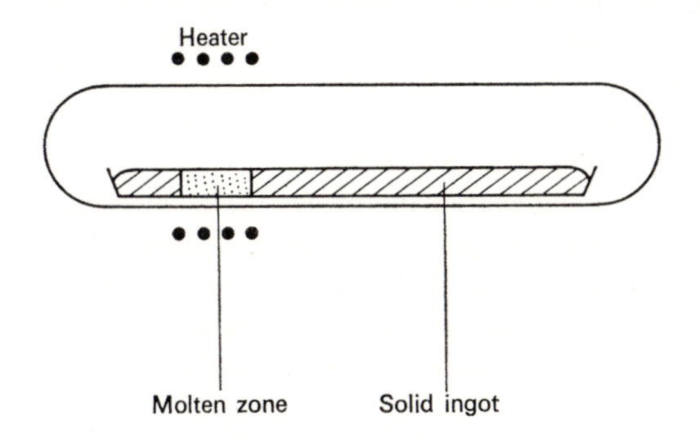

FIG. 2.1. Zone melting experimental arrangement.

portion and heavily contaminated extremities. The zone refining technique is described in much greater detail by Pfann[3] and some innovations on it by Brice.[4]

Before leaving this section on purification it should be noted that it is possible to purchase these elements from chemical manufacturers with impurity contents as low as 1 part in 10^5 or 10^6. However, unless extreme care is taken in handling the materials oxidation will occur and therefore before any attempt is made at compound preparation a short vacuum bake will be invaluable to remove the oxide layer. Container materials used by suppliers often leave much to be desired, for example, mercury of high purity supplied in polythene bottles soon picks up surface dust and ceases to be of high purity.

2.2. Phase Equilibria

In all of the II–VI systems a single compound only is formed with equal proportions of the two elements. The studies of the phase equilibria have provided information about the temperature–composition (T–x) section. In addition much useful information has resulted on the dissociation pressures of the II–VI compounds, a subject which has been reviewed by Goldfinger and Jeunehomme.[5] The growth of crystals from the melt and the vapour phase has placed the emphasis of investigations on solid–liquid and solid–vapour equilibria respectively. Heat treatment after crystal growth requires a knowledge of the solid state boundaries in the T–x section and the nature of the pressures that exist over the compounds.

In contrast to their constituent elements the II–VI compounds have high melting points and relatively low vapour pressures. Experimental evidence suggests that none of the compounds exists in the vapour species in detectable quantities. The metallic elements exist as monoatomic vapours and the semiconducting or insulating group VI elements appear in the main as diatomic vapours in equilibrium with the solid compound at elevated temperatures. The compound dissociation is thus described by a typical reaction for cadmium telluride, de Nobel[6]

$$\mathrm{CdTe(solid)} \rightleftharpoons \mathrm{Cd(vapour)} + \tfrac{1}{2}\mathrm{Te_2(vapour)}$$

If it is assumed that the activities of the elements are equal to their partial pressures then the equilibrium constant of the reaction is given by

$$K_{\mathrm{CdTe}} = p_{\mathrm{Cd}}p_{\mathrm{Te_2}}{}^{\frac{1}{2}} \tag{2.1}$$

The total pressure P over the solid CdTe is equal to the sum of the constituent elemental pressures

$$P = p_{\mathrm{Cd}} + p_{\mathrm{Te_2}} \tag{2.2}$$

At any given temperature the total pressure will change, if the pressure of one of the components is varied, and a minimum value P_{min} will be obtained for a particular p_{Cd} or $p_{\mathrm{Te_2}}$. Thus

$$P = P_{\mathrm{min}}, \text{ when } \frac{dP}{dp_{\mathrm{Cd}}} = \frac{dP}{dp_{\mathrm{Te_2}}} = 0$$

Combining equations (2.1) and (2.2)

$$P = p_{\mathrm{Cd}} + \frac{K^2_{\mathrm{CdTe}}}{p^2_{\mathrm{Cd}}} \qquad (2.3)$$

$$p_{\mathrm{Cd}} = 2^{1/3}\, K^{2/3}_{\mathrm{CdTe}} = 2p_{\mathrm{Te_2}} \qquad (2.4)$$

at the minimum and

$$P_{\mathrm{min}} = \frac{3}{2}\,(2^{1/3}\, K^{2/3}_{\mathrm{CdTe}}) = \frac{3}{2} p_{\mathrm{Cd}} \qquad (2.5)$$

The assumption that K_{CdTe} is constant at a given temperature is reasonable, provided the composition of the solid CdTe does not change appreciably. The P_{min} serves to indicate the pressure that is established when the solid is allowed to evaporate in a large vessel. Further, since $p_{\mathrm{Cd}} = 2\, p_{\mathrm{Te_2}}$ at the P_{min} line, the vapour has just the composition of the solid and so stoichiometric sublimation occurs at this condition. The suppression of sublimation is achieved by the application of an excess cadmium or tellurium pressure such that $p_{\mathrm{Cd}} \gg (p_{\mathrm{Cd}})_{\mathrm{min}}$ or $p_{\mathrm{Te_2}} \gg (p_{\mathrm{Te_2}})_{\mathrm{min}}$. The sublimation requires the diffusion of both components the rate of which is determined by the partial pressures. Clearly the smaller pressure of the two must determine this rate which, since equation (2.1) still holds in the presence of an applied pressure, is greatly diminished; this fact is of importance in melt and solution growth of the stoichiometric compound. Figure 2.2 illustrates the P_{min} or sublimation lines as derived by three different workers on the *P–T* diagram for the cadmium–tellurium system.[7] The experimental techniques which have been used to investigate the vapour pressures over two compounds include mass spectrometric analysis, effusion from a Knudsen cell[5] and optical density measurements.[11]

The *T–x* section of the Cd–Te system has been investigated in considerable detail and a maximum melting point at approximately the CdTe composition of 1092°C was obtained. Figure 2.3 illustrates the form of the liquidus that is derived from the combined results of de Nobel[6] and Lorenz.[8] Lorenz[7] has taken the Cd–Te system as quite typical of II–VI compounds. The peak in the liquidus near the maximum melting point suggests association of the constituents in the melt; on the other hand the sharp rise in the liquidus near the pure components indicates that the dissociation into constituent atoms or ions predominates. Thermal analysis studies in the region of the CdTe composition and similarly in the Hg–Te system seem to indicate that a range of solid solutions exists about the binary compound com-

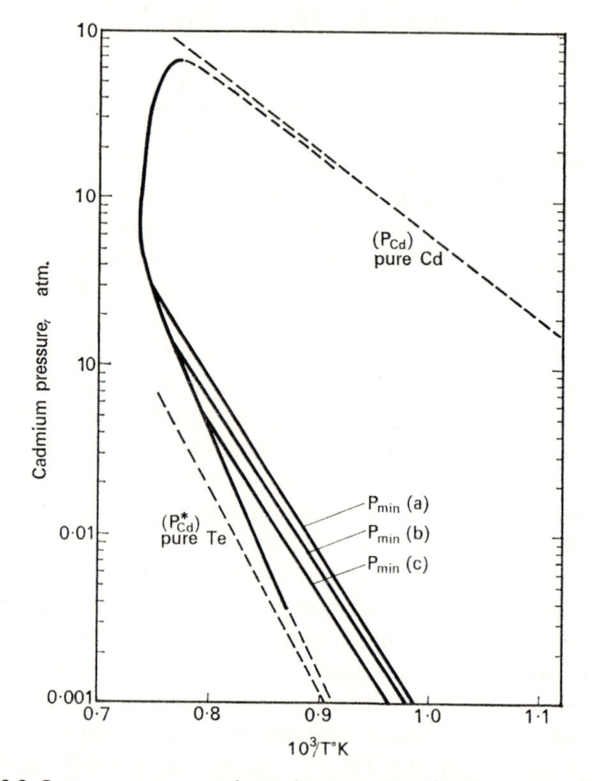

FIG. 2.2. $\text{Log}_{10}p_{Cd}$ versus reciprocal temperature for the system Cd–Te. $P^*_{Cd} = \sqrt{K_p/p_{Te2}}$, where the pressure of pure Te is inserted in the bracket. P_{min} references: (a) De Nobel,[6] (b) Lorenz,[8,9] (c) Brebrick and Strauss.[10]

positions.[6,10–12] For CdTe it is only of the order of 0·01 at.% of both Cd and Te in CdTe, but for HgTe, although there is not total agreement on the point, estimates of the solubility of Te in HgTe range from less than 0·1 to 1·5 at.%. Figure 2.4 compares the T–x sections for the Cd–Te and Hg–Te systems in the vicinity of the binary compound.[6, 12] Other systems where *T–x* sections have been studied are Zn–Te,[13, 14] Zn–Se,[7, 15] Cd–Se,[16, 17] Hg–Se[18] and to a very limited extent Cd–S.[19] The Cd–Se and Hg–Se investigations have an interesting occurrence in that a two-phase liquid region occurs on the selenium-rich side of the binary compound. Figure 2.5 shows the *T–x* section of the HgSe–Se system. It should be added that recently this two-phase liquid situation has also been suggested for the Hg–Te system.[20]

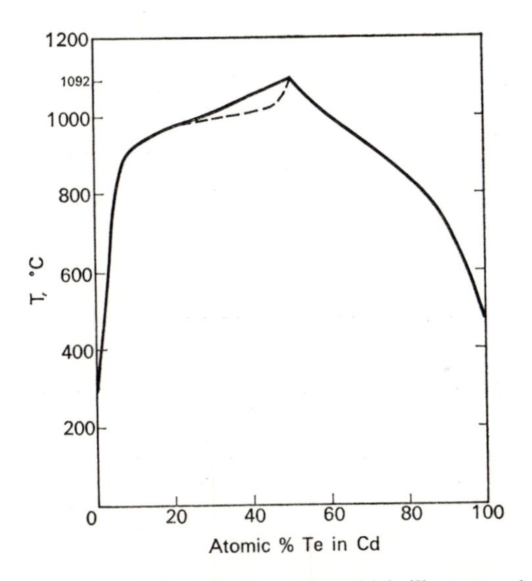

FIG. 2.3. *T–x* section of the Cd–Te system which illustrates the liquidus curve. The broken line represents the region in which the earlier results of de Nobel[6] differ from the final results of Lorenz.[8]

2.3. Preparation of the Compounds

Preparation by the direct reaction of the ultrapure elements is the technique which is used if at all possible. However, over half of the II–VI compounds of interest (ZnO, HgS, CdS, ZnS, CdSe, ZnSe) have extremely high melting points and in these materials some form of chemical reaction at room temperature is usually employed. A typical

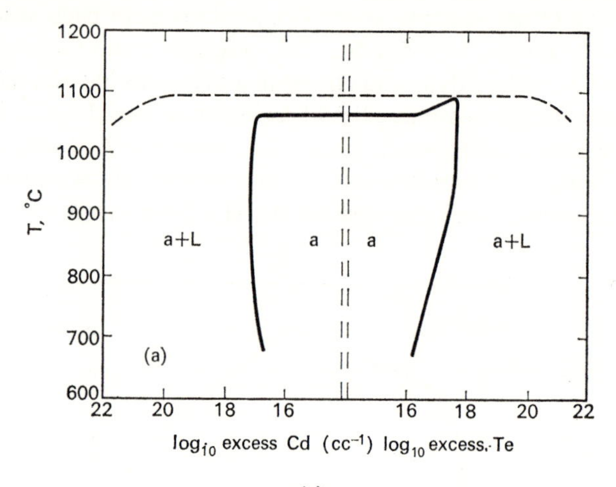

(a)

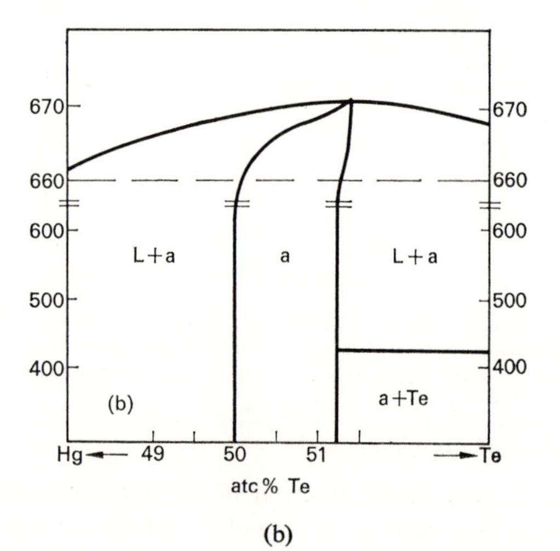

(b)

Fig. 2.4. Deviations from the stoichiometric compounds CdTe and HgTe: (a) CdTe, de Nobel,[6] (b) HgTe, Delves and Lewis.[12]

method of preparation is that used for the sulphide, in which the sulphide is precipitated from a solution of the metallic salt by hydrogen sulphide or an alkali sulphide. The remaining compounds (HgSe, ZnTe, CdTe, HgTe) have all been prepared by direct synthesis from the elements at elevated temperatures. In addition HgS, CdSe and ZnSe have been obtained by both techniques. However, as with elements the compounds can be obtained from chemical manufacturers with very high levels of purity and in some instances in single-crystal form.

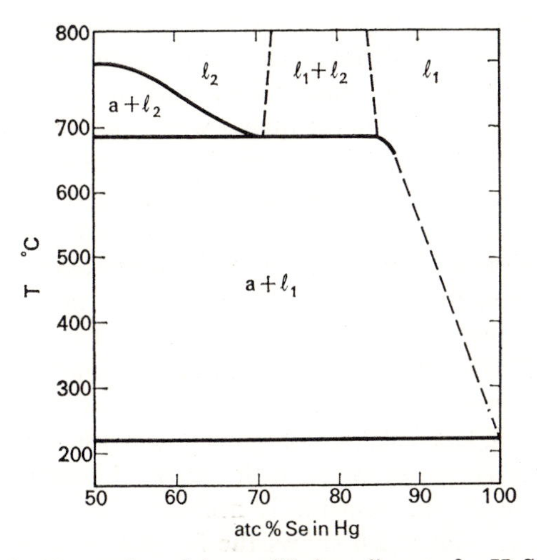

FIG. 2.5. The *T–x* section of the equilibrium diagram for HgSe–Se which illustrates the two-phase liquid region, Strauss and Farrell.[18]

In the preparation by direct synthesis the appropriate quantities of the purified elements are set in quartz or other inert tubes and sealed under vacua of the order of 1 microtorr. The tube is placed in an electric furnace at room temperature and the temperature is raised slowly to the point where the reaction occurs as determined from previous phase equilibria studies. The temperature rise is then continued more rapidly to some 50°C above the melting point of the compound. For congruently melting compounds rapid cooling does not affect the uniformity of the composition or the stoichiometry. However, for mercury telluride and to a lesser extent cadmium telluride the congruently melting point is on the tellurium-rich side of the stoichiometric compound. It is necessary to anneal at temperatures close to the solidus in order to obtain uniform composition, stoichiometric material. With CdTe it is possible to quench in the assumed uniform liquid phase and then anneal to solid equilibrium. HgTe, on the other hand, cannot be quenched because liquid mercury separates out, so that slow cooling to the anneal temperature is required and then after annealing the fall to room temperature needs 30–70 hr.[21] It is preferable when synthesis involves zinc or cadmium in the free state that the inside wall of the quartz tube is carbon coated. Both of these elements cause weakening of the quartz when they are used at temperatures in excess of 1000°C. The electrical properties of most of the compounds prepared by direct synthesis indicate that stoichiometric deviations are common. Giriat[22] has demonstrated the implications of the discussion in section 2.2 that the application of an external pressure of one of the constituent elements to a heated compound can significantly change its physical properties. He has obtained electron mobilities of 77,000 cm^2/V-sec at 77°K in HgTe which had been heated in a mercury atmosphere at 250°C.

Purification of less volatile compounds can usually be effected by zone refining *in vacuo*, however, with the II–VI compounds an excess pressure of one of the elements is necessary to prevent preferential constituent loss. Figure 2.6 shows the furnace arrangement that was employed to prepare CdTe by de Nobel.[6] De Nobel has, in fact, used the zone refining technique to dope the pure material with elements such as Cu, Ag, In and Sb. The foreign element is placed at one end of a boat that is filled with pure CdTe, the molten zone is then passed along the boat to and fro some 10 times to give a homogeneously doped ingot.

The high vapour pressures that are encountered with ZnSe, ZnS and CdS at their melting points render the zone refining technique impractical. Aven and Woodbury[23] have used the technique of solvent extraction to remove impurities. The compound which is surrounded by its molten constituent group II element is heated for a period of days. In this time approximately 5% of the compound is dissolved and many impurities are removed, particularly those of the acceptor type Cu, Ag, Au. When this method is applied to single crystals, intrinsic electrical properties are observed below room temperature. Extremely high-purity material of these large energy gap crystals is characterized by intrinsic behaviour down to very low temperatures. The temperatures used for this extraction are in the range 900–950°C. The extraction of chlorine and oxygen which are likely impurities in the sulphide and to a lesser extent the selenides is

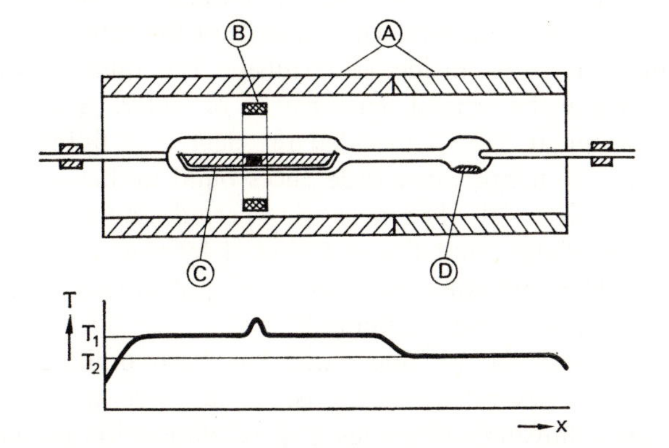

FIG. 2.6. Apparatus for zone purifying and zone levelling of CdTe. The material (C) is contained in a graphite boat enclosed in a silica container. A double nichrome furnace (A) and small molybdenum furnace (B) are used for heating. The cadmium pressure is produced by a cadmium reservoir (D).

achieved by prefiring the powder for a period of hours at 1000–1100°C in a stream of H_2S. The purity of the product may also be measured by the appearance of edge emission only, when the powder is excited by short wave-length radiation, and the absence of self-activated fluorescence. The doping of the sulphides and selenides requires for activator impurities the addition of the appropriate metallic nitrate or sulphate in dilute solution and the mixture to be fired in an H_2S atmosphere. Halogen coactivators are incorporated by firing the sulphide in the vapour stream of a mixture of H_2S and the halogen hydride.[24]

A technique which has been applied to tailor a material to particular requirements is that of hydrothermal synthesis. Kremheller and Levine[25] have used it to produce particle sizes of 5 microns or less in sulphide photoconductors so as to enhance the absorption of radiation. The small particle sizes result from a low preparation temperature, 350°C, which inhibits particle growth. The sulphide together with the salts, that contain the activator and coactivator impurities in the correct proportions, are sealed into a quartz vial partially filled with deionized water. The quartz vial is enclosed in a high pressure steel autoclave which also contains water. Since the liquid phase will be in equilibrium with the vapour, the pressure will depend on the temperature. Thus almost equal pressures inside and outside the quartz vial can be maintained below the critical temperature of water. Two phases are present for water at 350°C provided the degree of filling is between 20 and 50%. The temperature differences build up between the autoclave and the quartz vial. The processing period used for the sulphides is of the order of days and this is followed by slow cooling to room temperature. The technique has also been applied successfully to the production of alloys of the sulphides.[26]

In the discussion so far zinc oxide has been omitted despite its significant use in several applications. Zinc oxide can be prepared by any of the following techniques: (1) decomposition of the nitrate; (2) heating the chloride in steam; (3) heating the oxalate in air. However, if high-purity ZnO is required, oxidation of high-purity evaporated zinc is a reliable method. In this respect much work has been done on thin layers of zinc oxide, Heiland *et al.*[27] With their technique metallic zinc is evaporated onto a substrate of glass or quartz cooled to below 190°K. The oxidation is effected by heating the thin metallic layer to temperatures in the range 350 to 650°C in an oxygen atmosphere and ZnO layers 0·3 μm thick are produced. To obtain larger thicknesses of polycrystalline ZnO up to 30 μm metallic zinc is heated in air with a flame. The vapour is oxidized and deposits a layer on the substrate at a high temperature. Strong oxidation which is dependent on a high temperature flame produces white layers of ZnO, while weaker oxidation gives zinc-rich, yellow ZnO.

2.4. Single-crystal Preparation

2.4.1. *Vapour Phase Growth*

Growth from the vapour phase requires the simultaneous presence of both the group II and group VI elements in the gaseous form. The origin of the elemental vapour species may be the dissociation of the preformed compounds, the dissociation of gases stable at room temperature or directly from the elements. The division of vapour transport techniques into dynamic and static results from the use of open and closed systems respectively.

The dynamic methods are those in which the growth of crystals is continuously controlled by an external carrier gas, whether it is the hydride of the group VI element or an inert gas. The method was used by Frerichs(28) to produce the first appreciable sized crystals of CdS. The growth process occurs in a long quartz tube that is situated in a horizontal furnace and has two separate gas flows. A nitrogen or argon flow sweeps along the vapour of the metallic element, which is heated in a crucible within the system, and causes it to mix with a hydrogen sulphide or selenide flow. In the high-temperature region of the furnace the hydride decomposes and the sulphur or selenium is free to combine with the metal at the walls of the tube to give crystal growth. Platelet, needle and ribbon-like single crystals are produced by this technique and their form and size are dependent on the tube diameter, flow rate, growth temperature, temperature gradients and the metal vapour which has its pressure adjusted to a value of 100 torr usually. The method has been used successfully to prepare ZnS, CdS, HgS, CdSe, HgSe and CdTe. In the case of HgTe either single-crystal tellurium or polycrystalline material is produced.

Both Kremheller(29) and Fochs and Lunn(30), to mention only two of the many modified Frerichs techniques, have used the flow of a single inert gas over a silica boat which contains the already formed compound. Kremheller has passed a flow of helium through a quartz tube which contains a silica boat filled with high purity ZnS. Nucleation and crystal growth have been produced on a hollow cooled rod which is inserted into the furnace, Fig. 2.7. Fochs and Lunn have made a study of the nucleation sources suitable for the growth of thin CdS platelets. Both silica fibres and quartz crystals have been found satisfactory as nucleation sites. The experimental arrangement they have used to grow platelets on a silica fibre is shown in Fig. 2.8. The charge of CdS, which has previously been purified in an argon flow at 1075°C for 10 min, is placed in the quartz tube at the cold end. It is then transferred to the hot zone (1000°C) by a temporary rotation of the furnace through 90°. In the meantime the growth fibre has been held at a minimum temperature of 1000°C. The vapour of the CdS is carried by the argon flow at a rate of 0·25 l/min into the cooler region

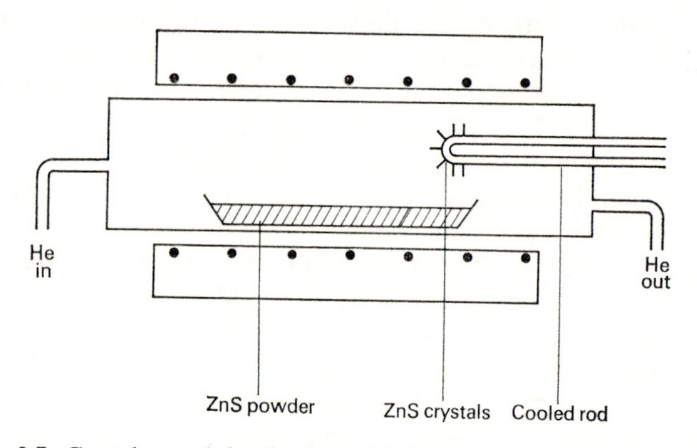

FIG. 2.7. Crystal growth by the flow of helium over ZnS and nucleation on a cooled rod.[29]

of the furnace and condenses on the walls as a crystalline deposit at temperatures in the range 950–700°C. The vapour flow is allowed to stabilize and the silica fibre is moved into the growth region, where nucleation on it commences. The whiskers that develop on the fibre after a time either thicken into needles or broaden out to form plates. The growth periods are 1–2 hr and the plates that result take three main crystal habits; these are the (10·0), (11·0) and (12·0) faces. When the silica fibre is replaced by a quartz crystal, fast growing (11·0) plates with a growth time of 1–2 min are frequently observed in addition to the slow growing plates mentioned above.

Dynamic flow methods have been applied to the growth of ZnO single crystals by a number of investigators.[31–34] Both Kubo[31] and Weaver[32] have used a flow of water vapour to hydrolyse a zinc halide. Weaver[32] has attempted to optimize conditions for growth of ZnO with the reaction zone maintained at 500°C and the water vapour swept across the $ZnCl_2$ by an argon carrier gas. A ratio of the water to zinc chloride vapour pressures in the reaction zone of between 4 and 7 produced yellow transparent crystals of ZnO with resistivities of 10^4 Ω-cm. The temperature of the growth region deter-

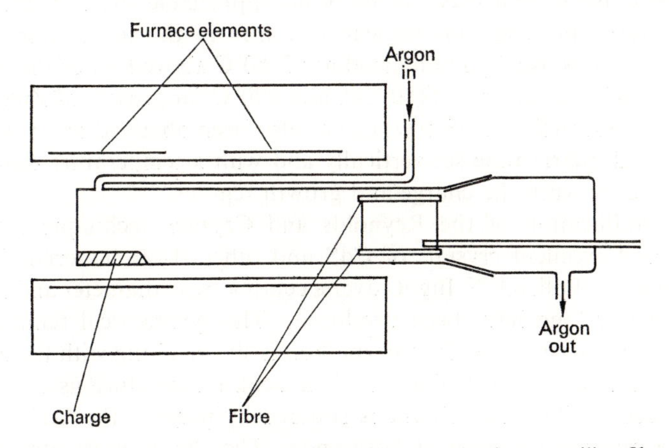

FIG. 2.8. Vapour growth of CdS single-crystal platelets on silica fibres in an argon flow.[30]

mined the crystal habit, at 1050°C both needles and platelets were obtained and the largest plates were about 4 mm across and 0·3 mm thick. Platelets of 1 cm^2 area have been produced by Takahashi[34] with a growth temperature of 1300°C. With this technique reduction in the growth temperature appears to favour the formation of hexagonal needles. Dodson and Savage[33] have employed a very different approach to ZnO crystal preparation by vapour transport in that sintered zinc oxide is used as the starting material. A hydrogen flow reduces the polycrystalline zinc oxide vapour to zinc vapour which is then re-oxidized to form single-crystal ZnO in a hot reaction zone. Hexagonal rods of ZnO up to 2 cm long and 3 mm in diameter have been grown at temperatures between 1280° and 1325°C. This latter technique is open to considerable further development particularly in terms of nucleation centres.

The static growth method revolves around the technique devised by Reynolds and Czyzack.[35] In this the powdered charge is placed in a quartz tube sited in the hot zone of a two zone furnace. A static atmosphere of purified H_2S or argon at a pressure of 200–300 torr is present in the quartz tube. Growth of crystals occurs after a period of about 100 hr and they can be of an appreciable size if provision is made for a predominant nucleation centre in the tube. The charge temperature is usually maintained at 50–80°C above that of the growth region and has values of 1180, 980 and 490°C for growth of ZnS, CdS, HgS.[36] Growth of ZnS crystals has also been obtained in a sealed off evacuated quartz tube set vertically and with a temperature difference of 100°C between the charge and growth regions.[37]

A modification of the Reynolds and Czyzack technique that has produced excellent crystals of CdS and other II–VI materials is that of Piper and Polich.[38] Ingots over a centimetre in diameter and several centimetres long have been produced. The symmetrical temperature profile that is necessary for this method is shown along with the experimental arrangements in Fig. 2.9. The peak temperature used for CdS was 1200°C. The CdS charge is sintered by packing the CdS powder into a quartz tube open at both ends. The charge is vacuum baked at 500–700°C for 1 hr, then it is fired in a stream of H_2S at 900–1000°C for 10 hr. This process purifies the charge and increases its density from 25 to 80% of that of the crystalline material. The quartz growth container has a blunt, conical end with a quartz rod joined to its tip to assist in the removal of the heat of sublimation. The charge at the other end of the growth chamber is backed up with a smoothly fitting closed quartz tube. The mullite tube which surrounds the quartz arrangement is vacuum baked at 500°C to remove volatile impurities. Argon is introduced into the mullite tube at a pressure of 1 atm and the furnace temperature is increased to the growth temperature. The growth chamber starts the process with its tip near the maximum temperature. The entire assembly is then moved at a rate of 0·3–1·5 mm/hr with the tip entering the cooler region first. Supersaturation at the tip increases until eventually nucleation commences; one growth seed usually crowds out all the others. The single crystal grows at the rate of movement of the assembly. In addition to CdS and ZnS subsequent experiments have shown that it is possible to grow CdSe, ZnSe, and ZnTe single crystals by this technique. Aven and Garwacki[39] have prepared

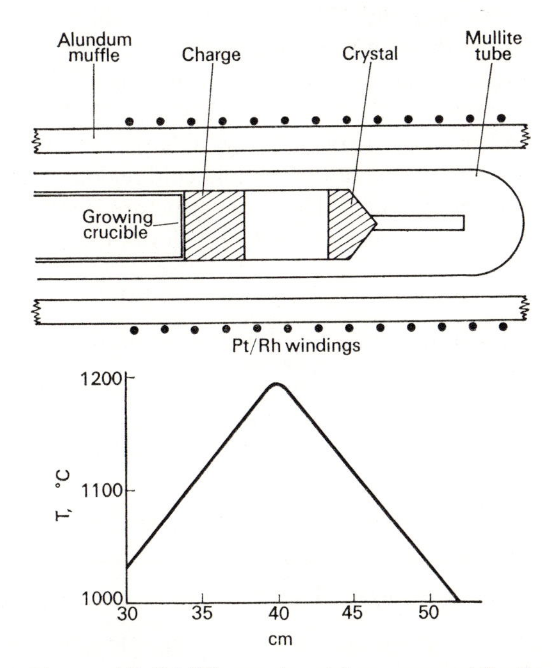

FIG. 2.9. Piper and Polich[38] experimental arrangement for the preparation of CdS single-crystal ingots by vapour transport.

solid solutions of the ZnSe–ZnTe system with complete uniformity along the length of the single-crystal ingot.

There is a further vapour transport method which has been applied to II–VI compounds by Nitsche *et al.*[40] and it involves the chemical transport of the metallic element by iodine. With this method it is possible to effect crystal growth at a much lower temperature. A small quantity of the halogen I_2 is mixed with the II–VI compound. If ZnS is taken as an example here, then the reversible reaction given below describes the process

$$ZnS + I_2 \rightleftharpoons ZnI_2 + \tfrac{1}{2}S_2$$

The polycrystalline mixture is set at the high temperature (1000°C) end of an evacuated silica tube and transport of the ZnI_2 and $\frac{1}{2}S_2$ commences to the cool end (750°C) where crystallization of the ZnS occurs. The iodine then diffuses back to the charge and recommences the transport cycle. It is found that the optimum conditions for growth are an iodine content of 5 mg/cm^3 which gives a ZnS transport rate of 20 mg/hr. After a period of 40 hr crystals of 5×5 mm^2 faces are obtained. The temperatures of the hot and cold regions are not critical, although as might be expected the crystal structure is influenced by the temperature of the cold region. Table 2.2 shows the results obtained by Nitsche.[40]

TABLE 2.2. PROPERTIES AND GROWTH TEMPERATURES OF II–VI COMPOUNDS PREPARED BY IODINE TRANSPORT

Compound	Colour	Crystal shape	Size (mm^3)	Crystal structure	T_1 (°C)	T_2 (°C)
ZnS	White	Polyhedra	$6 \times 6 \times 2$	Zinc blende	1000	750
ZnS	White	Hexagonal columns	$6 \times 4 \times 4$	Predominantly wurtzite	1300	1150
ZnSe	Yellow/Orange	Polyhedra	$4 \times 4 \times 3$	Zinc blende	1050	800
CdS	Orange	Hexagonal columns	$6 \times 6 \times 8$	Wurtzite	850	650
CdSe	Black	Hexagonal needle	$8 \times 2 \times 2$	Wurtzite	1000	500

It is evident that a certain amount of the iodine is incorporated in the crystals and influences their properties, as is shown by the work of Beun *et al.*[41] on CdS and Ballentyne and Ray[42] on ZnS.

2.4.2. *Melt Growth*

The II–VI compounds have not proved to be very amenable to growth from the melt. The main reason being that dissociation of the compounds in the region of their melting point occurs; this results in the preferential transport of one component to lower temperature regions. There exists also the problem of a suitable container material which will withstand the temperatures and pressures at the melting points of the II–VI compounds without being attacked by the metallic elements. Silica is the obvious choice as a refractory container since it can be obtained in a highly pure form; however it softens at 1300°C and thus only HgSe, HgTe, CdTe and ZnTe can be melt grown in silica containers with any assurance.

There appear to be future prospects in the system of liquid encapsulation for the melt growth of II–VI compounds. The principle of liquid encapsulation was first used in the growth of the volatile compounds, PbTe and PbSe.[(43)] In the liquid encapsulation technique an inert liquid is used which completely covers the melt. The loss of volatile components from the melt is prevented if an inert gas pressure is applied to the surface of the liquid encapsulant and is in excess of the equilibrium vapour pressure of the most volatile constituent in the melt. The requirements of the liquid encapsulant are that it be less dense than the melt, optically transparent and chemically stable in its environment. There should be no diffusion or convection of the volatile constituent through the layer, hence it is desirable that the volatile constituent should be insoluble in the liquid. Boric oxide (B_2O_3), barium chloride ($BaCl_2$), calcium chloride ($CaCl_2$) and $BaCl_2 + KCl$ are just some of the liquids which have been successfully used to grow single crystals of the III–V arsenides and phosphides.[(44)] The method of application of an encapsulant such as B_2O_3 is to cover the growth charge with B_2O_3 powder to a depth of 1 cm when it is molten. The melt requires a mirror clean surface. Growth of the single crystal on a seed ensues at the interface of the two liquids. The crystal as it is withdrawn from the liquid is covered by a thin film of B_2O_3 which is an important aspect of the technique since it reduces volatile component loss from the crystal. The B_2O_3 also wets the silica crucible and forms a liquid film between the crucible and the melt. Boric oxide has the additional property that it removes trace oxides and some impurities from the melt. At present there are no reports of the application of liquid encapsulation to II–VI compound growth.

The Stockbarger[45] method has been applied successfully to the growth of CdTe.[46,47] In the method molten material is contained in a carbonized silica tube which had been sealed off under vacuum. The

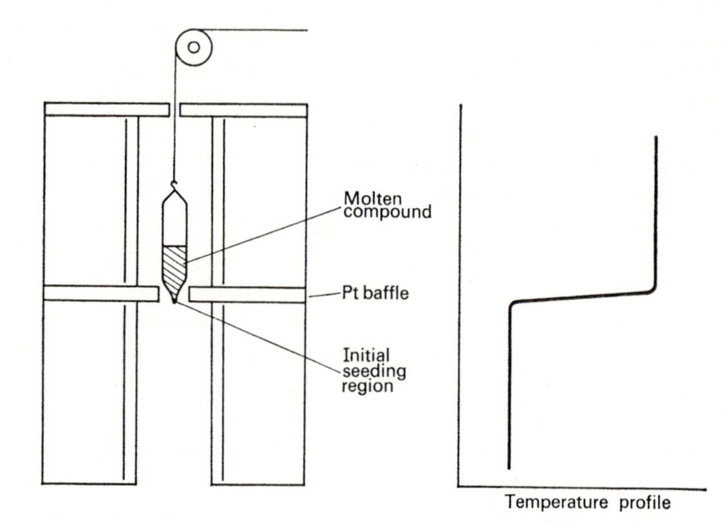

FIG. 2.10. Stockbarger arrangement for crystal growth.

silica tube is mounted vertically in a two zone furnace with a platinum baffle to separate the two regions which have a temperature difference of 50–80°C, Fig. 2.10. The silica tube has a constricted sharp end at the bottom to facilitate seeding in the growth process. The tube is lowered slowly through the temperature step at a rate of about 10 mm/hr and single-crystal growth ensues on the dominant growth seed. Wardzynski[48] has grown CdSe by this technique at a growth rate of 25 mm/hr.

Many refined modifications of the basic Stockbarger method have lead to crystal growth of other II–VI compounds. Harman and Strauss[49] have prepared both HgSe and HgSe–HgTe solid solutions in single-crystal form under a controlled mercury pressure with a growth rate of 4 mm/hr. In fact, it is possible to grow all tellurides in this manner. ZnTe which is normally obtained in *p*-type form has been produced in *n*-type form when prepared in a zinc atmosphere. The experimental arrangement involves a high pressure autoclave in which an inert gas

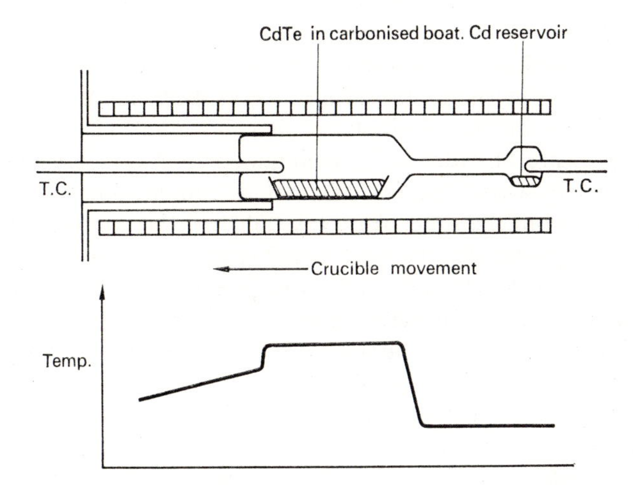

FIG. 2.11. Horizontal growth of CdTe single crystals.

pressure of 50 atm is maintained to constrain the zinc atmosphere around the ZnTe. The single-crystal *n*-type ZnTe which grows on passage through the temperature step is quenched to prevent inhomogeneity resulting from the surface diffusion of zinc.[50] ZnSe single crystals have also been grown from the melt under an applied argon pressure of 120 atm to suppress constituent vapour loss.[51] The maximum temperature reached in the ZnSe melt growth was 1700°C and necessitated the use of a vitreous carbon crucible.

Single-crystal preparation in the horizontal plane has been effected with a three-zone furnace arrangement which has the temperature profile shown in Fig. 2.11.[6] The carbonized silica boat which contains the CdTe has a pointed end to promote single-crystal growth. The boat is moved through the temperature step at 1 cm/hr and the crystal grows with departures from the stoichiometric composition that depend on the temperature of the cadmium reservoir. A variation of this method has been used by Lorenz[9] to investigate the growth of CdTe from the melt and also from a solution rich in one constituent. The technique is clearly applicable to most of the telluride and selenides and has been employed successfully to prepare HgTe.[52]

The zone melting technique has been extended to the preparation of the tellurides in single-crystal form. De Nobel[6] has prepared single-crystal CdTe with a set up similar to that shown in Fig. 2.6, in which a vapour pressure of cadmium is maintained from a reservoir. He has observed that when liquid CdTe has a composition which differs appreciably from the solid phase crystallizing, the component present in excess is segregated at the solid–liquid interface. Since both cadmium and tellurium have appreciable vapour pressures at the melting point of CdTe, the segregated excess leads to the formation of gas bubbles at the interface. If these bubbles are frozen as the growth proceeds then a highly porous ingot results. Thus an external cadmium pressure is necessary to overcome this effect and because of the tendency of CdTe to sublime, a rather larger pressure is applied to assist in the growth of uniform single crystals. The molten zone is moved at a speed of 5 mm/hr. The crystals obtained are found to grow preferentially along the [111] direction and have a tendency to twin in the (111) plane. The technique has been extended to the growth of HgTe and HgTe–CdTe solid solutions with only minor modifications being required.[46] Thick walled silica tubing is needed to encapsulate HgTe and its alloys with high proportions of mercury in order to prevent explosions.

Lorenz and Halsted[54] have developed a vertical zone refining technique to produce high-purity CdTe with electron mobilities of 0·105 and 5·7 m^2/V-sec at 300 and 20°K respectively. The refined ingot contains single-crystal grains of several centimetres length. In the technique they have simply used the solid mass of CdTe above the molten zone to prevent the loss of constituents. The elements are contained in a quartz tube that is coated with pyrolytic graphite. An inch long cylindrical quartz plug is inserted in the tube and rests just above the charge. The tube is then sealed off at the quartz plug under a vacuum of better than 1 microtorr. Reaction of the elements is carried out at 1150°C and the tube is then lowered to produce directional

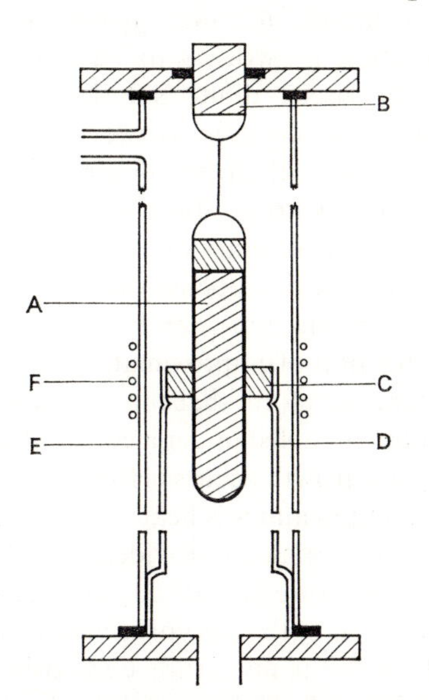

FIG. 2.12. Vertical zone refining assembly.[54] A. Quartz enveloped ingot. B. Shaft to variable speed drive mechanism. C. Graphite susceptor. D. Quartz susceptor holder. E. Quartz for protection from the atmosphere. F. Induction coil.

freezing of the ingot. It is arranged once more that the quartz plug is resting on the ingot and the tube is sealed off under vacuum. An induction coil and graphite susceptor are used to create a molten zone, which is passed from the tip of the quartz tube upwards by lowering the quartz tube. The whole process is effected in an atmosphere of nitrogen and the zone is at a temperature of 1130°C (40°C above the melting point of CdTe). After some twenty sweeps, at a rate of 2 cm/hr, high-purity material and crystals are obtained. Figure 2.12 illustrates the arrangement.

The above discussions on melt growth have been confined to almost stoichiometric conditions. In practice the alternative possibility of growth from solution in the II–VI compounds is not favoured. This is because of the form the temperature–composition equilibrium diagrams take. The liquidus rises steeply from the elementary components; therefore if the advantage of lower growth temperatures, that are associated with solution growth, is to be obtained then very dilute solutions have to be used; dilute solutions unfortunately produce small and inhomogeneous crystals. However, there are two pieces of experimental work on growth from solution that it is felt are worth drawing to the readers attention. Both methods have resulted from a good knowledge of the phase diagram.

In investigations of the Cd–Te phase diagram Lorenz[8] has observed that a maximum cadmium pressure P_{max} of 6·7 atm occurs at the liquidus for a composition of approximately 40 at. % cadmium. From Fig. 2.3 it can be seen, therefore, that if a solution which contains more than 40 at. % cadmium is cooled under a constant cadmium pressure less than P_{max}, CdTe crystal growth occurs. However, if the solution contains less than 40 at. % cadmium and it is heated under the same pressure, then CdTe crystal growth occurs again provided that the crystal is quenched before the remelt temperature is reached; thus a single crystal has been produced by heating. In both these processes a constant pressure line has been followed on the temperature–composition diagram.

Delves[20] has produced single-crystal HgTe from a tellurium-rich solution. The temperature–composition diagram for Hg–Te has a two-phase liquid region similar to that observed in Hg–Se and Cd–Se. The two liquid phases react monotectically with the HgTe phase at 664°C. A melt composition of $Hg_{1.0}Te_{1.125}$, which falls in the two-phase liquid region, has been used to give single crystals of HgTe with the monotectic solid composition in the initial growth section of the ingot. The technique, although it offers no practical advantage in the growth of HgTe, has real possibilities for the growth of HgSe and CdSe single crystals, since their monotectic temperatures are 100 and 350°C respectively lower than the melting points.

2.4.3. *Thin Film Growth*

Thin film behaviour in the II–VI compounds has provoked much interest recently with the advance towards smaller and smaller devices. Although films of compounds such as ZnS have been well known for many years, the requirements of high crystallinity had not hitherto been too important in their application. However, current trends show a more fundamental interest in the character of thin films and how they may be compared to bulk material. In this section it is intended that a brief account be given of the preparation of some II–VI thin films along with appropriate references.

The technique of vacuum deposition springs immediately to mind when thin films are discussed and is a subject that has been well reviewed by Holland.[55] Thin films of CdS and CdSe, two of the best known photoconductors, can be evaporated in conventional vacuum chambers onto heated substrates at pressures in the range 10^{-5}–10^{-6} torr. The source of the material has been either the separate heated elements or a charge of the preformed compound which is heated directly. It is found that the properties of the thin films of II–VI compounds differ greatly from those of a preformed compound; this is a direct result of the dissociation tendency of the II–VI materials. For some applications the change in the character of the thin films is not too critical provided that carrier mobilities and the resistivity remain high. Examples are the work on thin films of CdS with substrates heated to 200°C[56] and of CdSe with unheated substrates,[57] both of which are connected with thin film field effect devices.

Different treatments have been used to overcome the deviations from stoichiometry which result from the dissociation tendencies. Heat treatment in a gas ambient leads to increased homogeneity of defects or improvement in the stoichiometry if the gas atmosphere is suitable. In luminescent films of the sulphides an atmosphere of H_2S mixed with a small percentage of HCl helps to provide uniform doping. Thin films that are covered by a powder of the II–VI compound when heat treated take on the characteristics of the powder, hence uniform doping with copper and silver is possible by this technique. Hot wall evaporation techniques reduce the condensation of the volatile vapour species and thermalize the vapour to temperatures which approach that of the substrate. ZnS films have been prepared by a hot wall method[58] and more recently high mobility HgTe films have been produced similarly.[59]

Evaporation techniques with the various modifications suggested in the previous paragraph have also been at the root of some very interesting studies on epitaxial growth of II–VI compounds. Several crystalline substrates have been used and it has been demonstrated that by careful choice of temperature thin films of II–VI compounds can be grown with either of the two crystalline modifications. CdS hexagonal films are deposited on mica substrates which are heated to temperatures in excess of 200°C. However, a partly cubic film is obtained if evaporation is effected from a Knudsen cell onto a molybdenum cubic (200) face at 175°C.[60] CdS cubic single-crystal thin films have been grown on the (111) phosphorous faces of GaP while hexagonal films grow on the (111) gallium faces of GaP.[61] Weinstein[61] and Dima[62] have grown hexagonal CdTe films on the $(00\cdot\bar{1})$ sulphur faces of a CdS substrate. Shiojuri and Suito[63] have observed that hexagonal CdTe films of 500 Å thickness grow on (00·1) faces of a mica substrate at 380°C. However, if the substrate temperature is between 70 and 250°C the cubic modification grows with the [111] axis normal to the (00·1) face of the mica. At lower temperatures random orientations of both hexagonal and cubic modifications occur. A study of the epitaxial growth of ZnTe thin films on various ionic substrates has produced two suitable materials in CaF_2 and BaF_2.[64] Single-crystal films of cubic ZnTe have been grown along the [111] axis on freshly cleaved (111) faces of CaF_2 and BaF_2 at temperatures greater than 400 and 320°C respectively. One final example of the epitaxial growth in II–VI compounds is the work on CdTe–HgTe alloys which independently has found BaF_2 as a suitable substrate.[65] Freshly cleaved (111) faces of BaF_2 heated to 110–120°C give 400 to 1000 Å thick films of the alloys which are oriented in the same plane as the substrate.

Vapour phase reaction is another technique used for single-crystal growth which has been adapted to the growth of thin films.[66] The vapours of the constituent elements of the compound are allowed to react and form a film on a heated substrate. The nucleation of the reaction occurs over the surface of the substrate and films of the sulphides and ZnO have all been formed by the method. Figure 2.13 illustrates the experimental arrangement used by Studer and Cusano[67] to deposit ZnS on large area glass substrates. Film thicknesses in the range 1 to 50 μm have been achieved at growth rates of 0·5 to 1 μm/min.

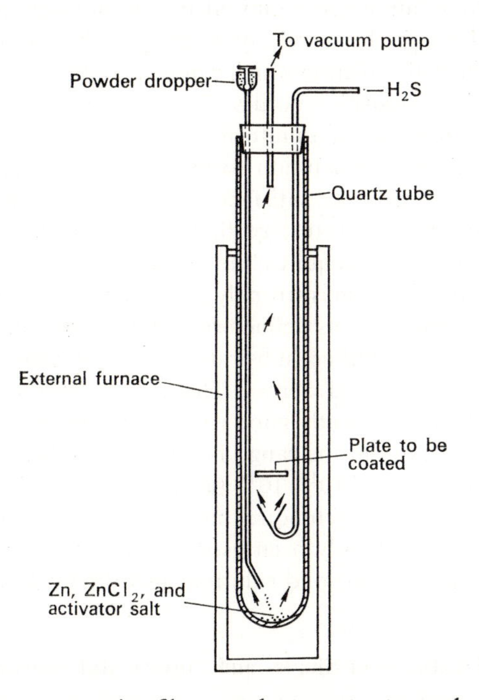

FIG. 2.13. Vapour reaction film growth apparatus to produce ZnS:Zn, Cl deposition on a flat substrate.[66]

An H_2S pressure of 0·5 torr on the substrate and a chamber at 500–650°C have produced the best results for ZnS. Cusano has extended this technique to the preparation of doped thin films of CdTe which he has used to fabricate solar cells. In a similar way the reaction of CdS films and vapours of $ZnCl_2$ has given the formation of ZnS–CdS solid solutions. Hexagonal ZnS epitaxial thin films have been grown by this method on CdS and GaAs single-crystal substrates.

A chemical spray deposition process has been developed principally to produce thin films of cadmium sulphide.[68] The process is quite simple in that the constituents are contained in aqueous solution of a metallic salt and an organic compound of the sulphur. The cadmium and sulphur in solution are atomized by the pressure of a gas which drives them through the nozzle of a spray. The sprayed mixture is incident on a heated substrate which can be maintained at temperatures up to 400°C while deposition occurs. The concentration of the solution determines the spray rate if a uniform film is to be obtained. Cadmium chloride and thiourea in solution form a complex which is a suitable starting material to produce satisfactory cadmium sulphide films. It is the decomposition temperature of the complex, in the example given above it is 210°C, that requires the use of a heated substrate. The crystallinity of the thin film is strongly dependent on the character of the substrate. Amorphous films defined by a grain size smaller than 400 Å are produced on crystalline substrates with the cadmium chloride–thiourea solution discussed above. The quality of films produced by spray deposition is sufficiently good to permit their use in photovoltaic cells. Film thicknesses in the range 1–1·5 μm have been observed with carrier concentrations of 10^{15}–10^{16} per cm^3 and mobilities as high as 90 cm^2/V-sec.[69] The technique has applications to many other compounds and has already been used to form the compounds CdSe, ZnS and ZnSe along with solid solutions of CdS–CdSe and CdS–ZnS.

2.5. Surface Treatment of Single- and Poly-crystal Specimens

It seems worthwhile to consider in this chapter the final treatment of specimens which have been cut or cleaved into a useful form. Most of the II–VI compounds are not hard and they may be easily polished with various grades of alumina powder. The resultant polished specimens have to varying degrees suffered surface damage which will cause the specimens to exhibit anomalous surface properties.

Etching techniques have been developed to remove the damaged surface layer and to produce a polished specimen of the required

TABLE 2.3. TABLE OF ETCHANTS AFTER GATOS AND LAVINE[70]

Material no.	Etchant	Conditions	Remarks
ZnS 28	0·5M $K_2Cr_2O_7$ in 16N H_2SO_4	Natural crystal 10 min 95°C	Etch pits on **A** surfaces
CdS 31	6 fuming HNO_3 6 CH_3COOH 1 H_2O	2 min	Sharply defined hexagonal pits on **B** surfaces
32	Same as 28	5–10 min 95°C	Polishing with dislocation etch pits on **B** surfaces Shallow dishing of **A** surfaces
33	1 H_2SO_4 100 H_2O 0·08 g Cr_2O_3	10 min 80°C	Difference in pit formation on **A** and **B** surfaces
ZnSe (ref. 71)	Concentrated NaOH solution	Hot	Polishes
CdSe 34	30 HNO_3 0·1 HCl 20 18N H_2SO_4	8s 40°C Rinsing in conc. H_2SO_4 to dissolve Se film	**A** surface develops hexagonal pits
35	1 HNO_3 3 HCl		Sharply bevelled pits on **A** surfaces
HgSe 40	50 HNO_3 10 CH_3COOH 1 HCl 20 18N H_2SO_4	10–15 min 40°C	Polishing
41	6 HCl 3 HNO_3 3 H_2O	Start with chemically polished surface 5 min. Film removed by immersing briefly in etchant No. 40 then brushing under H_2O; repeat	**B** surfaces develop triangular pits

(*see overleaf*)

TABLE 2.3. TABLE OF ETCHANTS AFTER GATOS AND LAVINE[70]—*contd.*

Material no.	Etchant	Conditions	Remarks
ZnTe 29	3 HF 2 H_2O_2 1 H_2O	2 min	Preferential, differentiates **A** and **B** surfaces
30	3 HNO_3 4 HF	8 s Rinse in conc. HCl, then dist. H_2O	Chemical polishing
CdTe 36	Same as 29	2 min	Etch pits on **B** surfaces longer times will polish **A** surfaces
37	"E" 10 ml HNO_3 20 ml H_2O 4 g $K_2Cr_2O_7$		Chemical polishing
38	"EAg-1" 10 ml E soln. 0·5–10 mg $AgNO_3$		Etch pit formation
39	7 saturated $K_2Cr_2O_7$ 3 H_2SO_4	2–3 min rinse in H_2O then in boiling 10% NaOH and $NaHSO_3$ then in H_2O and in ethylene diamine tetra-acetic acid	
HgTe 42	6 HNO_3 1 HCl 1 H_2O	10–15 min, 25°C	Polishing
43	1 HCl 1 HNO_3	Start with chemically polished surface, three 1 min etching with H_2O rinsing in between	Pits on **A** surfaces with background figures
CdTe 44 HgTe Alloys	6 HNO_3 1 HCl	Rinse in 1 HCl 1 CH_3OH	Polishes CdTe (0·05) HgTe (0·95) Results vary with composition

dimensions, which exhibits the same properties as the material before cutting to shape; such an etch merely acts as a polish. However, etches may be developed to act preferentially to show up dislocations or particular crystallite faces. A review of such etches has been made by Gatos and Lavine[70] for most known semiconductors and Table 2.3 shows those pertinent to II–VI materials. In the table references are made to **A** and **B** surface, **A** surfaces are those terminating in group II atoms and **B** surfaces in group VI atoms.

Sullivan and Bracht[72] have devised a method for surface treatment which combines the polishing and etching processes. A 30% HCl solution is fed continuously into the region between a polishing disc and the sample block. Extremely flat surfaces (nonplanarity of 0·5 μm) of CdS crystals have been obtained by the method on both the **A** and **B** surfaces normal to the *c*-axis. CdS surfaces prepared in this manner are excellent for the epitaxial growth of CdS films.

References

1. LAWSON, W. D., and NIELSON, S., *Preparation of Single Crystals*, pp. 18, 90, 141. Butterworth, London, 1958.
2. PFANN, W. G., *J. Metals* **4,** 747 and 861 (1952).
3. PFANN, W. G., *Zone Melting*. Wiley, New York, 1958.
4. BRICE, J. C., *The Growth of Crystals From the Melt*, pp. 129–34. North Holland, Amsterdam, 1965.
5. GOLDFINGER, P., and JEUNEHOMME, M., *Trans. Faraday Soc.* **59,** 2851–67 (1963).
6. DE NOBEL, D., *Philips Res. Repts*. **14,** 361–430 (1959).
7. LORENZ, M. R., *Physics and Chemistry of II–VI Compounds*, p. 84. North Holland, Amsterdam, 1967.
8. LORENZ, M. R., *J. Phys. Chem. Solids* **23,** 939–47 (1962).
9. LORENZ, M. R., *J. Phys. Chem. Solids* **23,** 1449–51 (1962).
10. BREBRICK, R. F., and STRAUSS, A. J., *J. Phys. Chem. Solids* **25,** 1441–5 (1964).
11. BREBRICK, R. F., and STRAUSS, A. J., *J. Phys. Chem. Solids* **26,** 989–1002 (1965).
12. DELVES, R. T., and LEWIS, B., *J. Phys. Chem. Solids* **24,** 549–56 (1963).
13. CARIDES, J., and FISCHER, A. G., *Solid State Comm.* **2,** 217–18 (1964).
14. REYNOLDS, R. A., STROUD, D. G., and STEVENSON, D. A., *J. Electrochem. Soc.* **114,** 1281–7 (1967).
15. WOSTEN, W. J., and GEERS, M. G., *J. Phys. Chem.* **66,** 1252–3 (1962).
16. REISMAN, A., BERKENBLIT, N., and WITZEN, M., *J. Phys. Chem. Solids* **66,** 2210–14 (1962).
17. BURMEISTER, R. A., and STEVENSON, D. A., *J. Electrochem. Soc.* **114,** 394–8 (1967).
18. STRAUSS, A. J., and FARRELL, L. B., *J. Inorg. Nucl. Chem.* **24,** 1211–13 (1962).

19. WOODBURY, H. H., *J. Phys. Chem. Solids* **24,** 881–4 (1963).
20. DELVES, R. T., *Brit. J. Appl. Phys.* **16,** 343–51 (1965).
21. SPENCER, P. M., *Brit. J. Appl. Phys.* **15,** 625–32 (1964).
22. GIRIAT, W., *Brit. J. Appl. Phys.* **15,** 151–6 (1964).
23. AVEN, M., and WOODBURY, H. H., *Appl. Phys. Letters* **1,** 53–54 (1962).
24. HOOGENSTRAATEN, W., *Philips Res. Repts.* **13,** 515–693 (1958).
25. KREMHELLER, A., and LEVINE, A. K., *J. Appl. Phys.* **28,** 748 (1957).
26. KREMHELLER, A., *et al.*, *J. Electrochem. Soc.* **107,** 12–15 (1960).
27. HEILAND, G., *et al.*, *Solid State Physics* **8,** 191–323, Academic Press, New York, 1959.
28. FRERICHS, R., *Phys. Rev.* **72,** 594–601 (1947).
29. KREMHELLER, A., *Sylv. Technol.* **8,** 11 (1955).
30. FOCHS, P. D., and LUNN, B., *J. Appl. Phys.* **34,** 1762–6 (1963).
31. KUBO, I., *Jap. J. Appl. Phys.* **4,** 225–6 (1965).
32. WEAVER, E. A., *J. Crystal Growth* **1,** 320–2 (1967).
33. DODSON, E. M., and SAVAGE, J. A., *J. Matl. Sci.* **3,** 19–25 (1968).
34. TAKAHASHI, T., EBINA, A., and KAMIYANA, A., *Jap. J. Appl. Phys.* **5,** 560–1 (1966).
35. REYNOLDS, D. C., and CZYZACK, S. J., *Phys. Rev.* **79,** 543–4 (1950).
36. HAMILTON, D. R., *Brit. J. Appl. Phys.* **9,** 103–5 (1958).
37. PIPER, W. W., *J. Chem. Phys.* **20,** 1343 (1952).
38. PIPER, W. W., and POLICH, S. J., *J. Appl. Phys.* **32,** 1278–9 (1961).
39. AVEN, M., and GARWACKI, W., *Appl. Phys. Letters* **5,** 160–2 (1964).
40. NITSCHE, R., BOLSTERLI, H. U., and LICHTENSTEIGER, M., *J. Phys. Chem. Solids* **21,** 199–205 (1961).
41. BEUN, J. A., NITSCHE, R., and BOLSTERLI, H. U., *Physica* **28,** 184–94 (1962).
42. BALLENTYNE, D. W. G., and RAY, B., *Brit. J. Appl. Phys.* **14,** 157–8 (1963).
43. METZ, E. P. A., MILLER, R. C., and MAZELSKY, R., *J. Appl. Phys.* **33,** 2016–17 (1962).
44. MULLIN, J. B., STRAUGHAN, B. W., and BRICKELL, W. S., *J. Phys. Chem. Solids* **26,** 782–4 (1965).
45. YAMADA, S., *J. Phys. Soc. Japan* **15,** 1940–4 (1960).
46. MCSKIMM, H. J., and THOMAS, D. G., *J. Appl. Phys.* **33,** 56–59 (1962).
47. INOUE, M., TERAMOTO, I., and TAKAYAMAGI, S., *J. Appl. Phys.* **33,** 2578–82 (1962),
48. WARDZYNSKI, W., *Proc. Roy. Soc.* **A260,** 370–8 (1961).
49. HARMAN, T. C., and STRAUSS, A. J., *J. Appl. Phys.* **32,** 2265–70 (1961).
50. FISCHER, A. G., CARIDES, J. N., and DRESNER, J., *Solid State Comm.* **2,** 157–9 (1964).
51. TSUJIMOTO, Y., ONODERA, Y., and FUKAI, M., *Jap. J. Appl. Phys.* **5,** 636 (1966).
52. WOJAS, J., *Phys. Stat. Sol.* **11,** 407–13 (1965).
53. LAWSON, W. D., *et al.*, *J. Phys. Chem. Solids* **9,** 325–9 (1959).
54. LORENZ, M. R., and HALSTED, R. E., *J. Electrochem. Soc.* **110,** 343–4 (1963).
55. HOLLAND, L., *The Vacuum Deposition of Thin Films*. Wiley, New York, 1960.
56. ZULEEG, R., and MULLER, R. S., *Solid State Electronics* **7,** 575–82 (1964).
57. WEIMER, P. K., *et al.*, *Proc. I.E.E.E.* **54,** 354–60 (1966).
58. KOLLER, L. R., and COGHILL, M. D., *J. Electrochem. Soc.* **107,** 973–6 (1960).
59. IGNATOWICZ, S., and KOBUS, A., *Bull. Acad. Pol. Sci.* **14,** 143–8 (1966).

60. ESCOFFERY, C. A., *J. Appl. Phys.* **35,** 2273–4 (1964).
61. WEINSTEIN, M., WOLFF, G. A., and DAS, B. N., *Appl. Phys. Letters* **6,** 73–75 (1965).
62. DIMA, I., and BORSAN, D., *Phys. Stat. Sol.* **23,** K113–15 (1967).
63. SHIOJURI, M., and SUITO, E., *J. Appl. Phys.* **3,** 314–19 (1964).
64. HOLT, D. B., *Brit. J. Appl. Phys.* **17,** 1395–9 (1966).
65. LUDEKE, R., and PAUL, W., *J. Appl. Phys.* **37,** 3499–501 (1966).
66. CUSANO, D. A., *Physics and Chemistry of II–VI Compounds*, p. 713 (M. Aven and J. S. Prener, eds.). North Holland, Amsterdam, 1967.
67. STUDER, F. J., CUSANO, D. A., and YOUNG, A. H., *J. Opt. Soc. Am.* **41,** 559 (1951).
68. CHAMBERLIN, R. R., and SKARMAN, J. S., *J. Electrochem. Soc.* **113,** 86–89 (1966).
69. CHAMBERLIN, R. R., and SKARMAN, J. S., *Solid State Electronics* **9,** 819–23 (1966).
70. GATOS, H. C., and LAVINE, M. C., *Progress in Semiconductors* **9,** 1–45, Heywood, London, 1965.
71. AVEN, M., MARPLE, D. T. F., and SEGALL, B., *J. Appl. Phys.* **32,** 2262–5 (1961).
72. SULLIVAN, M. V., and BRACHT, W. R., *J. Electrochem. Soc.* **144,** 295–7 (1967).

CHAPTER 3

FUNDAMENTAL OPTICAL PROPERTIES

IT IS proposed in this chapter to discuss absorption edge spectra, edge emission and dispersive properties in II–VI compounds. The data obtained from these optical properties along with other information derived from non-optical measurements will be related to the energy band structure in section 3.4. The consideration of non-edge emission (luminescence) and photoconductivity will be deferred until subsequent chapters.

3.1. The Absorption Edge

3.1.1. *General Features*

The spectrum in the region of the absorption edge exhibits much useful information about the material under investigation. A particular feature in the studies of II–VI compounds is the presence of exciton states which appear as peaks on the long wavelength side of the absorption edge. Excitons are bound electron–hole pairs which can be described in terms of "hydrogenic" levels close to the conduction band. These levels provide a series of discrete parabolic bands below E_c, the energy at the bottom of the conduction band, which is situated at $\mathbf{k} = 0$ in the II–VI compounds. A reduced mass μ is used in the equation of the parabola and for cubic structures this is related to the electron and hole effective masses by $\mu^{-1} = m_e^{*-1} + m_h^{*-1}$. In the wurtzite structure the strongly anisotropic nature of the effective mass gives a more complex expression for μ.

The absorption of radiation that leads to electronic transitions between the valence and conduction bands is split into direct and indirect processes. Direct transitions require that in the excitation process no change in the **k**-value of the electron occurs. Indirect transitions represent the condition that there is a change in crystal momentum (**k**) of the electron. Thus vertical (direct) transitions are important when the valence and conduction band extrema are located at the same point in the Brillouin zone. For indirect transitions the band extrema differ in their position in **k**-space. Figure 3.1 illustrates both types of transition in an example where the indirect transition is important. The indirect absorption process requires the annihilation or creation of phonons to balance the crystal momentum. Experimentally it is possible to differentiate between direct and indirect

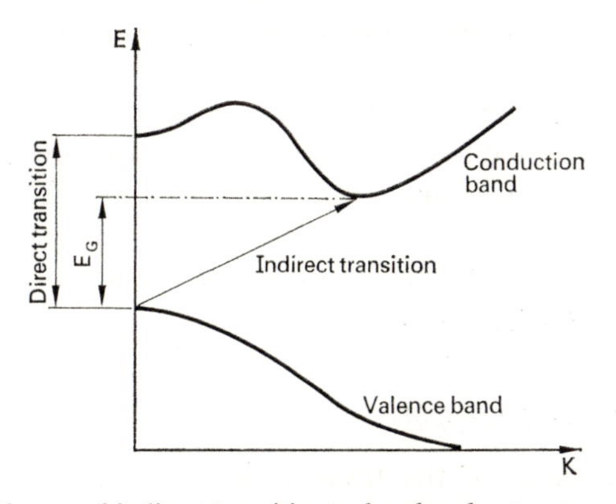

FIG. 3.1. Direct and indirect transitions when band extrema are located at different positions in **k**-space.

processes by the level of the absorption coefficient α; α takes values from 10^4 to 10^5 cm^{-1} for the former and 10 to 10^3 cm^{-1} for the latter at the absorption edge. Further α has an energy dependence of $E^{\frac{1}{2}}$ for direct transitions and of E^2 for indirect transitions.

The modifications that are imposed on the edge spectrum by the exciton states appear in low-temperature studies. The peaks which occur on the long wavelength side of the absorption edge give an indication of the binding energy of the exciton state. For the zinc blende structure spherical energy surfaces are assumed and the exciton binding energy E_{be} is related to the depth of different exciton states below the conduction band by $E_n = -E_{be}/n^2$; n takes values 1, 2, 3, ..., for the ground and various excited states. The variation of the exciton energy with crystal momentum is parabolic as mentioned in the first paragraph and the exciton's momentum is determined from the sum of the electron and hole effective masses. The energy dependence of the absorption coefficient for transitions to exciton states is similar to that for band to band transitions.

Free carriers provide a further influence on the form of the edge spectrum. If there is an appreciable concentration of free carriers in the conduction or valence band, then the general level of absorption rises. This in turn creates difficulties about the edge location which is generally based on an absorption coefficient of 10 cm^{-1}. In degenerate materials the lowest energy states in the conduction band are filled and thus the energy required to excite electrons is increased, i.e. the absorption edge is shifted to shorter wavelengths. Impure HgSe and HgTe both exhibit this effect and since their electron effective masses are low, they have a small effective density of states in the conduction band. As a result early work on these two compounds led to spurious values for their energy gaps.

3.1.2. *Experimental Methods*

The measurement of the transmission or reflectivity of a sample provides a satisfactory way to determine the form of the absorption edge. The relation which exists between the transmission T and the absorption coefficient[(1)] is

$$T = (1 - R)^2 \exp(-\alpha d)$$

where R is the normal reflectivity and d the sample thickness. R is in turn related to the refractive index n and the extinction coefficient

$$k = \frac{\alpha\lambda}{4\pi} \qquad \text{by } R = \frac{(n-1)^2 + k^2}{(n+1)^2 + k^2}$$

which abbreviates to

$$R = \frac{(n-1)^2}{(n+1)^2}$$

Since $\lambda \sim 10^{-5}$ cm at the absorption edge in the II–VI compounds with the exception of HgSe and HgTe, then k is much less than one. Hence in the analysis of transmission measurements, the reflectivity can be calculated from the refractive index if it is known. Alternatively, the $(1 - R)$ term is eliminated, if specimens of more than one thickness are studied. There exists a practical limitation in transmission measurements, in that the detection of radiation does not go below a certain level. Thus with large absorption coefficients thin specimens (10^{-4} cm) are required either in the form of thin evaporated films or as thin platelets such as may be grown for CdS. The thickness limitation on samples, which are taken from a single-crystal ingot and polished, is 10^{-3} cm and sets a limit to the absorption coefficient of 10^4 cm^{-1}. It is necessary to consider very much smaller spectral variations of reflectivity to obtain α through the Kramers–Kronig relation at the high levels of absorption observed with exciton transitions. The reflectivity can be used to provide information about the effective mass, if the free carrier concentration is known.

Details of the location of the energy bands in k-space may be determined from the pressure dependence of the absorption edge. An empirical approach compares the known dependencies for the group IV elements, Ge and Si. These are as follows for the conduction minima of Ge with the valence band maximum at [000]: 12×10^{-6} eV/kg-cm^{-2} for the [000] minimum, 5×10^{-6} eV/kg-cm^{-2} for the [111] minimum and -2×10^{-6} eV/kg-cm^{-2} for the [100] minimum. These results coupled with the level of absorption coefficient will help considerably in the identification of the semiconductor energy band configuration. Additional information about the shape of the energy bands is derived from magneto-optical effects and cyclotron resonance measurements. These and other experimental details are described more fully by Moss,[1] Lax and Mavroides.[2] Optical reflectance modulation technique is a powerful and recent tool, which has been applied generally to the optical investigations of semiconductors.[3] It is particularly important in the accentuation of effects at high symmetry points. Both electroreflectance and thermoreflectance have been used in the study of the fundamental absorption edge in CdS and CdSe with considerable success.[4, 5]

3.1.3. *Experimental Results*

It is only by taking an overall view of results from the different experimental methods that an assessment of the nature of the absorption edge can be made realistically. Segall and Marple[6] discuss fully the intrinsic absorption with special emphasis on excitons for the sulphides, selenides and tellurides of zinc and cadmium. Their collected data coupled with work in other references cited here[7–11] will suffice to

TABLE 3.1. OPTICAL ENERGY GAPS OF II–VI COMPOUNDS AT LOW TEMPERATURES (W—WURTZITE STRUCTURE, Z—ZINC BLENDE STRUCTURE)

Compound	E_g eV	$\frac{dE_g}{dT}$ 10^{-4}eV/°K
ZnO	3·44 W	−9·5
ZnS	3·91 W	−8·5
	3·84 ZB	−4·6
CdS	2·58 W	−5·2
HgS	2·10 ZB	
ZnSe	2·80 W	
	2·83 ZB	−8
CdSe	1·84 W	−4·6
HgSe	−0·1 ZB	
ZnTe	2·39 ZB	−5·0
CdTe	1·60 ZB	−2·3
HgTe	−0·1 ZB	

provide a satisfactory coverage of the edge absorption in II–VI compounds. The low temperature values of the energy gap or absorption edge energy as determined from transmission and reflectivity measurements are listed in Table 3.1.

The detailed reflection spectrum at low temperatures shows two or three exciton peaks dependent upon whether the crystal structure is zinc blende or wurtzite. The peaks result from the splitting of the valence band through spin–orbit interaction and crystalline field effects as was intimated in section 1.4. The Γ_8 uppermost valence band in zinc blende is separated by Δ_{so} from the Γ_7 band. In wurtzite the quadruply spin degenerate Γ_5 band is split into Γ_9 and Γ_7 bands with a separation related to the crystalline field energy Δ_{cr}. In addition to this the lowest valence band with Γ_7 symmetry is Δ_{so} lower in energy than the uppermost Γ_8 band. Exciton absorptions are designated by superscripts A, B, and C to indicate the transitions from the three valence bands in order of increasing energy. It is possible therefore to derive Δ_{so} and Δ_{cr} from the exciton spectrum in wurtzite and Δ_{so} in zinc blende where

TABLE 3.2. EXCITON ABSORPTION AND BINDING ENERGIES AND VALENCE BAND-SPLITTING ENERGIES (IN ELECTRON VOLTS)[6, 12, 13]

Compound	Structure	E_{ex}^{A}	E_{ex}^{B}	E_{ex}^{C}	$E_{be}^{A} = E_{be}^{C}$	E_{be}^{B}	Δ_{so}	Δ_{cr}
ZnO	W	3·777	3·383	3·422	0·059	—	0·0087	—
ZnS	ZB	3·799	—	3·871	0·0401	—	0·072	—
	W	3·871	3·900	3·990	0·0401	0·0403	0·092	0·055
CdS	W	2·554	2·569	2·632	0·0294	0·0295	0·065	0·027
ZnSe	ZB	2·799	—	—	0·019	—	0·43	—
	W	2·860	2·876	2·926	—	—	—	0·061
CdSe	W	1·826	1·838	1·851	0·0157	0·0167	0·42	0·041
ZnTe	ZB	2·381	—	—	0·010–0·013	—	0·9	—
CdTe	ZB	1·596	—	—	0·010	—	0·9	—

only the A and C exciton peaks are observed. At liquid helium temperatures the exciton absorption to the first excited state as well as the ground state is observed. The separations between the Γ_9 level and the two Γ_7 levels are denoted by E' and E'' and are related to the crystalline field and spin-orbit splitting energies by [2]

$$E'('') = \tfrac{1}{2}(\Delta_{so} + \Delta_{cr})(\mp)[\tfrac{1}{4}(\Delta_{so} + \Delta_{cr})^2 - \tfrac{2}{3}\Delta_{so}\Delta_{cr}]^{\frac{1}{2}}$$

Table 3.2. lists the A, B, and C exciton ground state absorption energies along with the exciton binding energies and splitting energies for several of the II–VI compounds.

Reynolds and co-workers[13, 14] have reviewed the optical properties of the II–VI compounds with particular reference to excitons and they include a detailed section on exciton theory.

The pressure dependence of the absorption edge has been studied in most of the II–VI compounds.[10, 11, 15, 16] Similar studies on germanium and silicon have permitted the determination of the lowest lying conduction band minimum. The pressure coefficients $(dE_G/dP)_T$ which characterized the principal conduction band minima in Ge were 12×10^{-6} eV/kg-cm^{-2} for the [000] minimum, -2×10^{-6} eV/kg-cm^{-2} for the [100] minimum. Madelung[17] has concluded from theoretical considerations that the general outline of band structure is the same in the III–V compounds as in the group IV elements. The experimental results seem to bear out these conclusions generally and only in two examples InAs and InP was there difficulty in distinguishing between a [111] and [000] minima. The same view is taken of the band structure in the II–VI compounds with the zinc blende structure and also for the wurtzite structure with certain obvious limitations.

The experimental pressure coefficients tend to be lower in the II–VI compounds than those in germanium. For the zinc blende structures 6×10^{-6} eV/kg-cm^{-2} is a good average value for $(dE_G/dP)_T$. With CdTe two values have been obtained; these are $4{\cdot}4 \times 10^{-6}$ eV/kg-cm^{-2} and 8×10^{-6} eV/kg-cm^{-2};[6] the latter value has been associated with an electron transition at the Γ point. The high absorption coefficients derived from the transmission data tend to suggest that the transition is direct. Further, the discussion on exciton edge absorption indicates the direct transitions occur at the Γ point. Both ZnSe and ZnTe change from a blue to a red shift at high pressures, a fact which points to [100] minimum predominating in this region. The wurtzite structures have lower pressure coefficients, the values for CdS and CdSe are approximately 3×10^{-6} eV/kg-cm^{-2}. In ZnS $(dE_G/dP)_T$ is 9×10^{-6} eV/kg-cm^{-2} and that in ZnO has values ranging from $(0{\cdot}6–2{\cdot}78) \times 10^{-6}$ eV/kg-cm^{-2}.[11, 12] ZnO changes from the wurtzite to zinc blende structure at 130,000 kg/cm^2 with an unaltered pressure coefficient. CdS and CdSe transform to the zinc blende structure and reverse the sign of their pressure coefficient to characterize the [100] conduction band minima. The wurtzite structures with the exception of ZnO are taken to be direct gap at the Γ point, although the pressure coefficients in themselves are not wholly conclusive. Alloys formed between wurtzite and zinc blende compounds of the II–VI series show a linear variation of energy gap with composition. This may reasonably be taken as an indication that the transitions are from the same conduction band minima at the extremes of a given alloy system. Hence the conduction band minima of the zinc blende or wurtzite compounds with the exception of ZnO appear to be at the same position in **k**-space. Zinc oxide, however, has a very low value of pressure coefficient; if the series ZnTe, ZnSe, ZnS, ZnO is studied, it is seen that the energy gap increases with decreasing atomic weight except for ZnO, which

TABLE 3.3. PRESSURE COEFFICIENTS AND CONDUCTION BAND MINIMA

Compound	Structure	(dE_G/dP) 10^{-6} eV/kg-cm^{-2}	Pressure range kg-cm^{-2}	Minimum assigned to pressure coeff.
ZnO	W	0·6–2·78	<130,000	[00·1]
	ZB	1·9	> 130,000	[111]
ZnS	W	9	180,000	[000]
	ZB	5·7–6·3	180,000	[000]
ZnSe	ZB	0·6	< 130,000	[000]
	ZB	− 2·0	> 130,000	[100]
ZnTe	ZB	6	< 40,000	[000]
		− 1	> 40,000	[100]
CdS	W	3·3	< 27,500	[000] or [00·1]
	ZB	− 0·7	> 60,000	[100]
CdSe	W	3·7	< 30,000	[000] or [00·1]
	ZB	− 1·5	> 30,000	[100]
CdTe	ZB	4·4–8	35,000	[000]

on an electronegativity calculation should have an energy gap at 0°K of 4·8 eV. On this evidence it would seem that the lowest conduction band minimum and the highest valence band minimum do not coincide in **k**-space for ZnO. However, the observations of direct exciton production would tend to suggest that the conduction and valence band extrema do coincide in **k**-space.[18] Table 3.3 lists the pressure coefficients and the location of conduction band minima based simply on these values.

Figures 3.2 and 3.3 illustrate two forms of absorption edge versus pressure curves. Figure 3.2 represents the variation in CdS for which a discontinuity occurs at 27,500 kg-cm^{-2} and also a change in the pressure coefficient. Figure 3.3 shows a similar plot for ZnSe in which only a

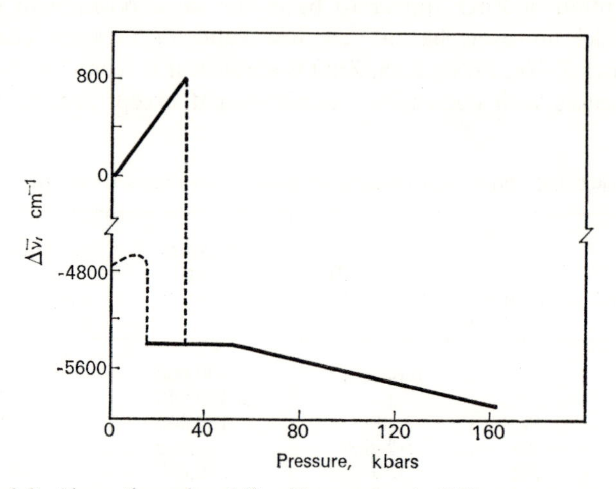

FIG. 3.2. Absorption edge shift with pressure in CdS at room temperature.[10]

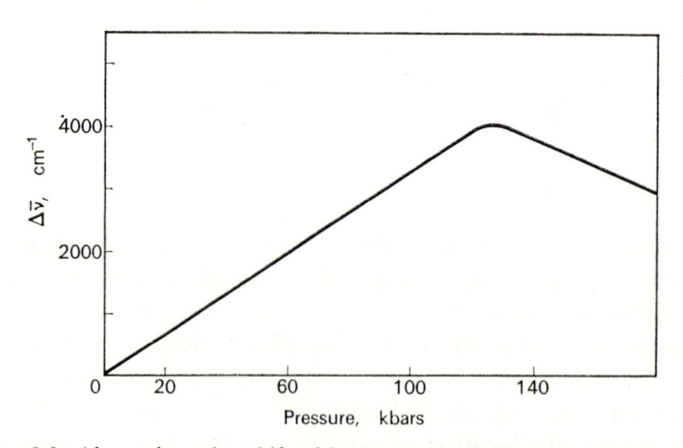

FIG. 3.3. Absorption edge shift with pressure in ZnSe at room temperature.[8]

change in pressure coefficient occurs at 130,000 kg-cm^{-2}. Figure 3.4 represents the energy band form of the zinc blende and wurtzite structures which are centred about the Γ point. The separation of the Γ_8–Γ_7 valence bands is shown for zinc blende. The wurtzite diagram indicates the transitions which occur for polarized radiation along with the two Γ_9–Γ_7 valence band separations.

In the foregoing discussion the band structure of mercury compounds has been omitted. The reasons for this are perhaps obvious since HgSe and HgTe have negative energy gaps, which effectively means free carrier effects dominate the absorption spectrum except at very

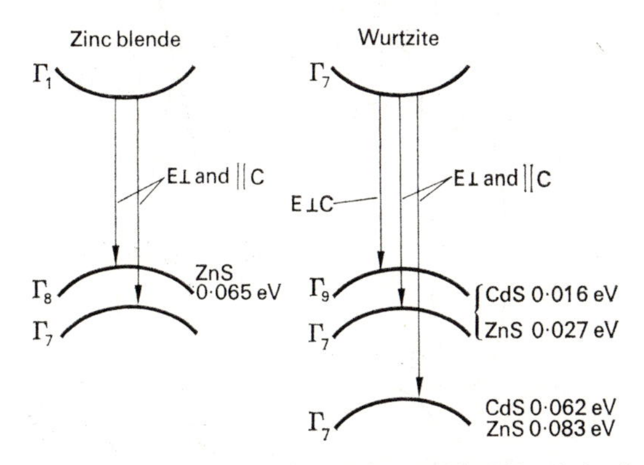

FIG. 3.4. Energy band structure of wurtzite and zinc blende in the region of the Γ point with crystalline field and spin-orbit splitting effects present in the valence band structure.[7]

low temperatures. It does, however, seem appropriate at this point to put forward two models for the band structure in HgSe and HgTe. Groves and Paul[19, 20] have formulated a band structure for HgTe, which is similar to that observed in gray tin. In this model the conduction and valence band extrema touch at the Γ point and the valence band maximum at a Λ-point is 0·1 eV above the Γ-conduction band minimum. The energy band structure in HgSe is likely to be similar to that in HgTe. A slightly different form has been suggested for HgTe on the basis of the alloying properties.[21, 22] HgTe, when alloyed with a compound such as CdTe or In_2Te_3, changes from semimetallic to semiconducting behaviour. Harman[21] has made an intensive experimental study of $Hg_{1-x}Cd_xTe$ and has observed the effective mass to decrease with x to almost zero at $x = 0{\cdot}17$. The conduction and valence bands were taken to touch at that composition and then to separate at compositions on either side of it. Figure 3·5 illustrates the variations in the band structure with composition and gives a possible form for it in HgTe. Very little consistent evidence is available on the behaviour of the absorption edge in HgS.

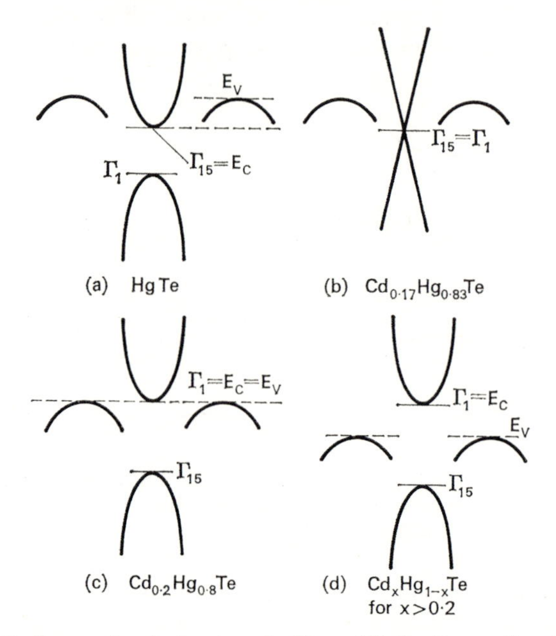

FIG. 3.5. Energy band structure in $Hg_{1-x}Cd_xTe$ alloys, after Harman *et al.*[21]

3.2. Lattice Absorption

The absorption of energy by lattice vibrations falls into a very different region of the spectrum (medium infrared) to that used in the absorption edge measurements in II–VI compounds. The lattice vibrations (phonons) in a periodic lattice are treated as harmonic oscillators, which are coupled to their nearest neighbour oscillations. Both transverse and longitudinal modes of oscillation with different velocities transmit the energy across the crystal. In crystal structures with more than one atom per unit cell there exists the possibility that adjacent atoms can oscillate in and out of phase with each other. The in and out of phase oscillations are termed acoustic and optical vibrations respectively. This leads to a classification of phonons into four classes which are:

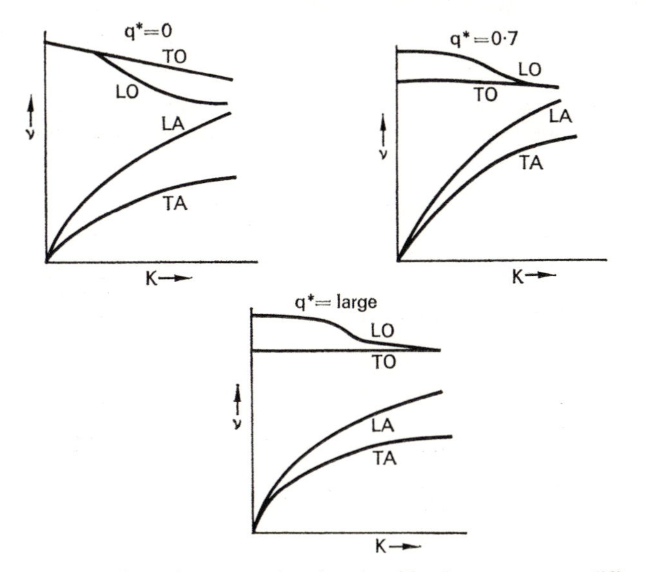

FIG. 3.6. The dispersion curve for the zinc blende structure at different values of ionicity q^*.[24]

longitudinal optical (LO), transverse optical (TO), longitudina acoustic (LA) and transverse acoustic (TA). An excellent discussion on the subject of vibrational modes in solids is given by Wang.[23] The optical modes are associated with higher energies since they result from out of phase motion. If the two atoms in the primitive unit cell are of the same type, the LO and LA phonon energies are equal at the zone edge. Unit cells, in which atoms of different masses exist, a separation occurs between the lowest optical frequency and the highest acoustic frequency. Mitra[24, 25] has extended the study of phonon frequencies ω as a function of wave vector $\mathbf{k}$ to take into account the effect of ionic character in the chemical bond. The ω–$\mathbf{k}$ dispersion curves, that he has derived for different values of ionicity q^* in the zinc blende structure, are illustrated in Fig. 3.6.

Information about the phonon energy is derived from both the reflectivity spectrum and the emission spectrum and is discussed below. Restrahlen bands (narrow absorption bands for radiation) in the reflectivity spectra of crystals permit the determination of the optical phonon

TABLE 3.4. TWO AND THREE PHONON ABSORPTION PEAKS IN ZnSe DETERMINED FROM THE FOUR CHARACTERISTIC PHONON FREQUENCIES[14]

LO = 0·0258 eV, TO = 0·0263 eV,
LA = 0·0201 eV, TA = 0·0108 eV

Peak position[25]		Assignment	Calculated position cm^{-1}
eV	cm^{-1}		
0·031	250	LA + TA	249
0·037	298	LO + TA	297
0·046	371	LO + LA	372
		TO + LA	
0·052	420	LO + TO	420
0·062	501	2LO + TA	503
0·073	588	2TO + LA	586

energies. Balkanski and Besson[26] in their studies of CdS observed peaks which were assigned to phonon coupling in the crystalline field. The interpretation of these and other results[27, 28] on CdS produced conflicting assignments to the different peaks in terms of multiple phonon absorption and emission. Marshall and Mitra[29] extended the above measurements to frequencies close to the transverse optical phonon frequency of 241 cm^{-1} determined by Collins.[30] They were able to interpret all the results in terms of six phonon energies. The two extra phonon energies were caused by the anisotropic nature of the crystalline field in the wurtzite structure and result in the splitting of transverse optical and acoustic phonon energies.[14] Similar studies have been made on wurtzite-structured ZnS and zinc blende-structured ZnS and ZnSe. Table 3.4 lists the peaks observed in the infrared spectrum of ZnSe[31] and the assignments given to them.

In the edge emission spectrum of many II–VI compounds a series of peaks on the long wavelength side of the edge is observed. These peaks have an equal energy separation which has the value of the longitudinal optical phonon energy. The LO phonon energy is related to the

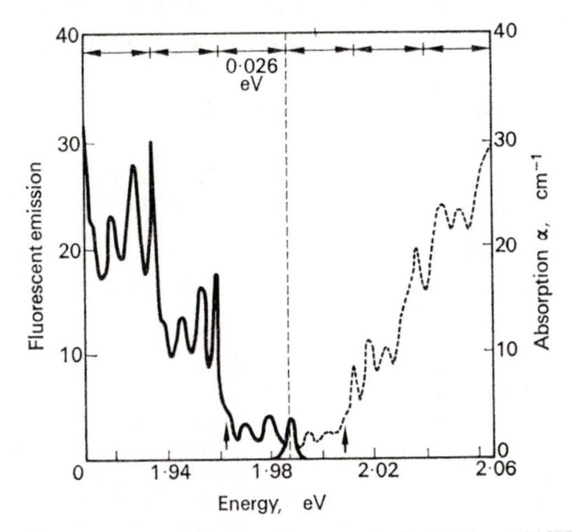

FIG. 3.7. The emission and absorption spectra in ZnTe at 20°K mirrored about a central no-phonon line at the intrinsic edge. The LO phonon energy interval of 0·026 eV is indicated at the top of the diagram.[32]

TO phonon energy by $E_{LO} = E_{TO}(\varepsilon_{\infty}/\varepsilon_S)^{\frac{1}{2}}$; ε_S is the static dielectric constant and ε_{∞} is the optical or high frequency dielectric constant. This represents, then, another method of determining the phonon energies. Within the LO phonon repeat interval there is considerable detail which has led investigators to identify peaks with the other phonons. A particularly interesting example is that obtained for ZnTe[32] in which the absorption and emission spectra form a mirror image about the intrinsic absorption edge, Fig. 3.7. These edge emission observations have been made on ZnO, ZnS, ZnSe, ZnTe, CdS, and CdTe. Table 3.5 summarizes the results obtained on phonon energies at the zone centre and the zone edge.

It is noticeable in this table that the phonon energies decrease with increasing atomic weight. Other features which are not so obvious are the dependence of $(LO/TO)^2$ on M_1/M_2, the mass ratio of the elements in the compound, and on the ionicity squared.[29]

Balkanski[35] in a review of phonon–phonon interaction gives a much more complex breakdown of phonon energies. In his classification of CdS he considers the phonon energies with respect to each Γ-point that arises along with the appropriate dispersion curve.

TABLE 3.5. PHONON ENERGIES IN ELECTRON VOLTS[29–34]

		Zone centre	Zone edge		Zone centre	Zone edge	Zone edge		
Compound	Structure	TO	TO_1	TO_2	LO	LO	LA	TA_1	TA_2
ZnO	W	—	0·054	0·052	—	0·060	0·030	0·017	0·012
ZnS	W	—	0·039	0·037	—	0·042	0·022	0·011	0·009
ZnS	ZB	0·29	0·037	—	0·031	0·026	0·020	0·011	—
ZnSe	ZB	0·026	0·026	—	0·031	0·026	0·020	0·011	—
ZnTe	ZB	0·024	0·022	—	0·026	0·023	0·016	0·006	—
CdS	W	0·030	0·030	0·030	0·038	0·037	0·019	0·010	0·009
CdSe	W	0·023	0·019	—	0·027	0·025	0·014	0·007	—
CdTe	ZB	0·0174	0·017	—	0·021	0·022	0·013	0·008	—

3.3. Edge Emission

Edge emission describes the radiative recombination processes which occur within a fraction of an electron volt of the absorption edge. The observation of these processes in any detail requires that liquid hydrogen and liquid helium temperatures are accessible to the investigator. It is, however, possible at higher temperatures to derive information about phonon energies as has been described in section 3.2. A very full discussion on this subject has been given by Halsted[34] in which he suggests that the extensive results obtained for CdS may provide a framework for the explanation of edge emission spectra of other II–VI compounds. The edge emission spectra in II–VI compounds is attributed to several processes, which include defect interaction, free excitons and bound excitons.

3.3.1. *Intrinsic and Defect Emission*

The intrinsic emission spectrum is associated with the recombination of electrons from the conduction band with holes in the valence band. Emission of this type is not observable in the II–VI compounds except possibly at elevated temperatures when exciton and defect states are ionized.

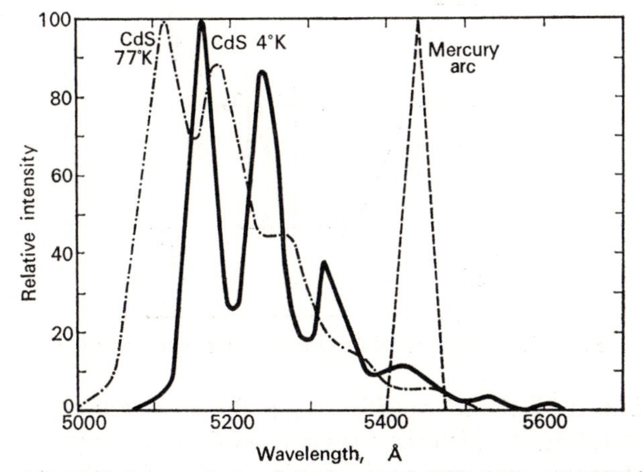

FIG. 3.8. Emission spectrum of single-crystal CdS at 77 and 4·2°K.[36]

The general structure of the edge emission spectrum at 4°K is a series of peaks separated by the longitudinal optical phonon energy with a decrease in peak intensity as the wavelength increases. Figure 3.8 illustrates the emission spectra from single-crystal CdS at 77 and 4·2°K.[36] A further feature of interest in Fig. 3.8 is that the emission energy of the zero phonon peak has shifted to shorter wavelengths as the temperature is increased. Pedrotti and Reynolds[37] have put forward an explanation for the observed shift on the basis of a detailed analysis of emission spectra from CdS at 4·2, 27, 33, 38 and 48°K. They assume that a bound electron state exists in the vicinity of a trapped hole. At 77°K the bound electron state is ionized and recombination occurs between a free electron and a trapped hole. At 4·2°K the recombination takes place between the two bound states which are separated by a smaller energy than one free state and one bound state. The energy difference between the two recombination processes is the binding energy of the electron. The energy model which describes this fluorescence in CdS is shown in Fig. 3.9.[37] The probability of process G_1 is low at 4·2°K and the G_5 transition gives the prominent

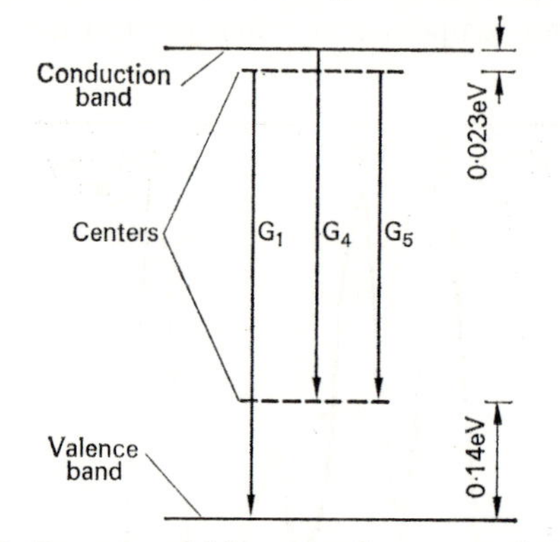

FIG. 3.9. Energy model for green fluorescence in CdS.[37]

peak, while at 77°K the transition G_4 is the dominant emission peak. The temperature at which the change from G_5 to G_4 transitions as the dominant process depends on the doping of the CdS. Broad confirmation of this model has come from time-resolved edge emission spectra at a range of temperatures as determined by Colbow.[38] He also observed additional structure in the two emission bands which was attributed to transverse optical phonons. Maeda[39] and Van Doorn[40] have shown that *n*-type conducting crystals have the G_5 low energy emission dominating to much higher temperatures, whereas *n*-type insulating crystals have only the G_4 high energy emission at all temperatures. Such crystals are produced simply by anneals at 600°C in cadmium vapour for conducting CdS and in vacuum for insulating CdS.[41]

A similar model has been used to describe the emission in the complex spectra of the predominantly cubic ZnSe[42]; the behaviour has been associated with stacking faults which result from the presence of small amounts of the wurtzite structure. In ZnSe the electron trapping centre is located 0·015 eV below the conduction band edge and the hole trapping centre 0·12 eV above the valence band edge. Investigations of CdSe and CdS–CdSe solid solutions at 77 and 4·2°K have indicated similar behaviour of edge emission energy with temperature. The edge emission spectra in CdSe are of a broader character than in CdS and the energy difference between the doublet emission peaks is only 0·009 eV.

Samelson and Lempicki[45] have studied the edge emission in cubic ZnS:Cl crystals from 10 to 300°K. The crystals studied were characterized by different stacking sequences, I-normal cubic and II-reversed cubic faults. Type I crystals exhibited two series of peaks and type II crystals only showed the longer wavelength series of peaks. The short wavelength peaks result from exciton transitions and will be considered further in section 3.3.2. The explanation of the long wavelength peaks has been proposed in terms of free hole transitions to a sulphur vacancy level (0·17 eV below the conduction band) and a chlorine substitution level (0·22 eV).

In investigations of the edge emission spectra in CdTe single crystals, Halsted and Segall[46] have observed a very similar spectrum to that in CdS and ZnSe. The temperature dependent doublet structure is observed at low temperatures in CdTe and these investigators have described the edge emission process by a transition to a doubly ionized acceptor level some 0·056 eV below the conduction band. They attributed the long wavelength peak of the doublet to an exciton bound to the ionized acceptor level. It was further suggested that this model could also be applied to the doublet structure in CdS. However, the donor–acceptor model of Pedrotti and Reynolds[37] would seem to explain most of the features of the doublet emission in II–VI compounds. The edge emission spectra associated with donor–acceptor transitions are characterized by broad peaks much greater than kT

in width. With increase in temperature these peaks broaden further and eventually destroy the longitudinal optical phonon structure. The broad peaks result from the range of separation of donor–acceptor pairs and the variety of possible defects interacting as pairs which occur in the host lattice.

3.3.2. *Exciton Emission*

(a) *Free excitons.* The consideration of free exciton transitions at the absorption edge has been dealt with in some detail in section 3.1. The free exciton is an excited state in a crystalline solid of an electron and a hole in orbit about each other at distances large compared with atomic dimensions. The observation of free exciton edge emission is made at temperatures well above that of liquid helium, where the exciton is usually part of a bound complex. The ionization energy of the bound complex to give a free exciton is generally much smaller than the ionization energy of the free exciton, hence the observation of free exciton emission at intermediate temperatures.

Polarization effects of the exciton edge emission relative to the c-axis of the wurtzite-structured CdS have been observed. Emission polarized normal to the c-axis is found to be independent of the nature of the incident radiation.[47] The temperature dependence of the ratio of emission intensity with $E \perp c$ to that with $E \| c$ gives an activation energy of 0·014 eV. This value agrees very well with that obtained by other methods for the crystalline field splitting of the Γ_5 valence band into Γ_9 and Γ_7 symmetries.[48] These results assume that the free exciton state has the same symmetry Γ_7 as the conduction band at $\mathbf{k} = 0$. Similar depolarization effects have been observed in the edge emission of ZnO,[49] ZnSe,[42] ZnS.[45] Free excitons move with a range of kinetic energies and their emission peaks have half widths of the order of 10^{-3} eV, which compares with a half width of 10^{-4} eV for the stationary bound exciton states.

(b) *Bound excitons.* Excitons have a tendency to form bound exciton complexes with natural defects which are a common feature in crystals of the II–VI compounds. The bound exciton complexes are created by the optical excitation of the crystals. The emission spectra which result from the disintegration of these bound exciton complexes occur on the long wavelength side of the free exciton emission spectra that are sometimes present. The extensive studies of Thomas and Hopfield[50] on CdS have resulted in the development of a theory of bound excitons. The theory is based on the band symmetry properties associated with the wurtzite structure of CdS. Four bound exciton complexes are postulated and these are excitons bound to neutral and ionized donor and acceptor states. The experimental evidence supports these ideas although one of the proposed complexes an exciton bound to an ionized acceptor has not yet been identified.

The different types of bound exciton complexes have been analysed from the magneto-optical behaviour of the emission spectra (Zeeman effect). The characterization of the three observed bound exciton conditions associated with the topmost valence band is given below.

(1) An exciton bound to a neutral acceptor exhibits emission that splits into a doublet when the magnetic field **H** is perpendicular ($\perp$) to the *c*-axis of the CdS crystal. The doublet separation indicates an electron *g*-value of $-1{\cdot}76$. At arbitrary orientations the line splits into a quartet and is indicative of anisotropic *g*-values. The line splitting depends linearly on the magnetic field which suggests a single unpaired electronic charge. The intensity ratio of the doublets is unaffected by temperature variation and hence a zero *g*-value for the ground state occurs.

(2) An exciton bound to a neutral donor shows a linear dependence of the emission peak energy on the magnetic field. With $\mathbf{H} \perp c$ the emission splits into a doublet and the magnitude of the splitting gives an electron *g*-value of $-1{\cdot}76$. In the orientation $\mathbf{H} \| c$ no splitting is observed, hence the *g*-value of the excited and ground states are equal. Temperature dependence of the emission intensities of the doublet confirm that an exciton is bound to a neutral donor.

(3) An exciton bound to an ionized donor has only been observed in the absorption spectrum. Zero field doublet splitting occurs although only the high energy component is observed. Both components are only seen for magnetic fields in excess of 10 kG and it is necessary to extrapolate back to obtain the zero field splitting of 0·0003 eV. The zero

field splitting arises from an exchange interaction of an unpaired electron and an unpaired hole in the upper state and characterizes an exciton bound to an ionized defect. Emission intensities of the doublet are independent of temperature and indicate a singlet ground state (ionized donor or acceptor). The small binding energy of the exciton to the ionized defect distinguishes the ionized donor from the ionized acceptor defect.

The binding energies of the excitons to the centres observed in CdS are for A excitons: (1) neutral acceptor—0·0177 eV, (2) neutral donor—0·0066 eV, (3) ionized donor—0·0038 eV; and for B excitons: (1) neutral acceptor—0·0099 eV, (2) neutral donor—0·0061 eV. A bound exciton complex has also been observed in ZnTe[32] which takes the cubic zinc blende structure. The free exciton has a binding energy of approximately 0·02 eV and is in turn bound to an ionized donor or acceptor level located some 0·4 eV below the conduction band minimum. Zero field splitting of the emission occurs at 4·2°K with an energy of 0·0017 eV. In the wurtzite structured CdSe the dissociation energy of an exciton bound to a neutral acceptor has been determined from magneto–optical measurements to be 0·008 eV.[51]

An additional feature of the bound exciton complex is a vibrational spectrum which appears as a fine structure similar to the vibration rotation spectrum of a singly ionized hydrogen molecule. Figure 3.10 illustrates the molecular-like series observed in CdS at 1·2°K[52] and represents the frequency difference for each line which extrapolates to the Γ_6 free exciton line.[53]

Halsted and Aven[54] have used the information gained in the magneto–optical investigations of CdS to identify spectra in other II–VI compounds. The effective mass ratio m_p^*/m_n^* in the II–VI compounds does not vary greatly and has an average value of approximately 4. They have taken the ratio of the dissociation energy of the bound exciton complex to the ionization energy of a neutral acceptor to be 0·10 and the ratio for a neutral donor to be 0·20. Bound exciton spectra have been identified on this criterion in several II–VI compounds for both neutral acceptors and neutral donors. Figure 3.11 illustrates the plot of dissociation energy of the complex versus ionization energy of the donor defect which seems to confirm the assertions made above.[54]

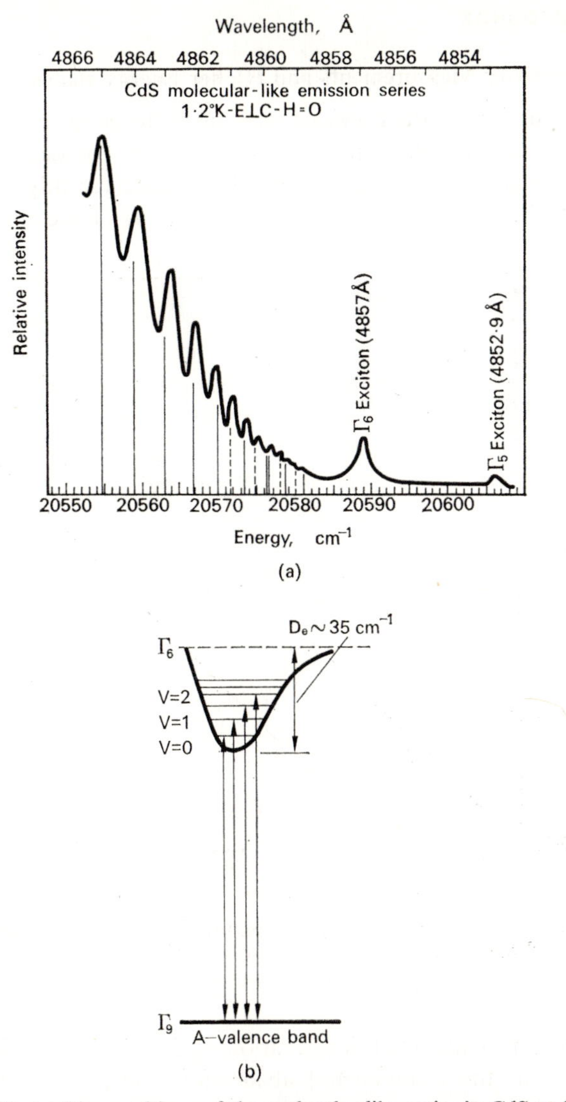

FIG. 3.10. (a) Line positions of the molecular like series in CdS at 1·2°K. The free exciton lines with Γ_6 and Γ_5 symmetries are also shown.[52] (b) Energy representation of the vibrational spectrum relative to the Γ_9 valence band and the Γ_6 free exciton level.[53]

3.4. Reflectivity Measurements and Higher Energy Bands

The discussions in the previous sections of the chapter have indicated the usefulness of reflectivity measurements in the determination of band gap and lattice vibration energies. More recently the study of reflectivity in the visible and ultraviolet regions of the spectrum have

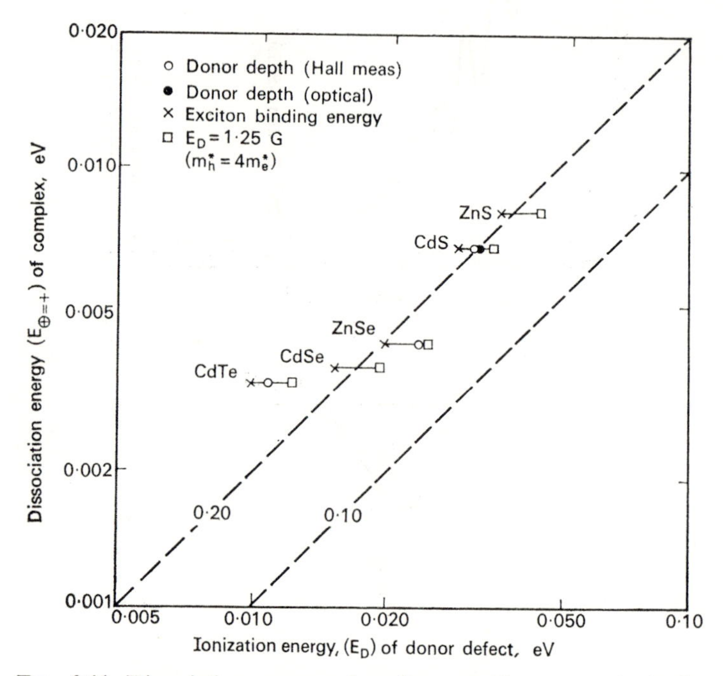

FIG. 3.11. Dissociation energy of exciton complex versus ionization energy of the donor defect in fluorescent emission spectra at 4°K for II–VI compounds.[54]

been used to provide much information on interband transitions apart from those at the fundamental absorption edge. These transitions, loosely spoken of as between higher energy bands, occur generally at symmetry points different from $\mathbf{k} = 0$ (Γ). The reflection spectrum of a crystalline solid exhibits anomalies at frequencies which correspond to resonance absorptions in that solid, such as would occur for an interband transition. Most of the higher energy band transitions in the II–VI compounds occur in the ultraviolet spectral regions; absorption spectroscopy is extremely difficult in these regions and reflectivity studies became almost indispensible as a means of study of these higher energy bands. The reflection and absorption spectra of the II–VI compounds at low temperatures in the region of the absorption edge have been characterized by considerable exciton structure both in ground and excited states. The experimental evidence of section 3.1 suggests that II–VI compounds generally have a direct band gap and the intrinsic excitons associated with the absorption edge are thought to be direct also. In a similar way it might be expected that direct exciton structure would occur at higher energy bands although the probability of excitons in excited states will be small because of the low occupation probability of these bands.

The higher energy band structure of II–VI compounds with zinc blende and wurtzite structures has been investigated by Cardona and other workers.[24, 31, 55–62]. The measurements have been made on single crystal samples, freshly cleaved faces obtained from bulk material or epitaxial layers at room and liquid nitrogen temperatures. Comparisons have been made between the spectra observed in compounds, which dependent upon the growth conditions, take either the zinc blende or the wurtzite structure. Table 1.5 illustrates broadly the transitions to be expected in the zinc blende and the wurtzite structures with the electric vector **E** of the radiation either parallel or perpendicular to the hexagonal *c*-axis.

3.4.1. *Zinc Blende Structure*

Cardona[55] studied the transmission properties of epitaxial layers of HgTe, HgSe, CdTe, ZnTe, ZnSe deposited onto quartz substrates for energies up to 5·5 eV at 80°K and room temperature. The only observable peaks were for the E_1 and E_1 plus the spin-orbit energy Δ_1 transitions. The E_1 ($\Lambda_3 \rightarrow \Lambda_1$ transition) values showed a similar trend to the fundamental absorption edge across this range of compounds while the spin-orbit splitting depended on the anion (Te or Se), which is associated with the valence band. Reflectivity studies[56] on cleaved surfaces of the three tellurides over the energy range 1·0–25 eV at room temperature and 1·0–6·5 eV at 77°K provided confirmation of the above results along with much extensive information on other transitions. A small additional peak was observed on the low

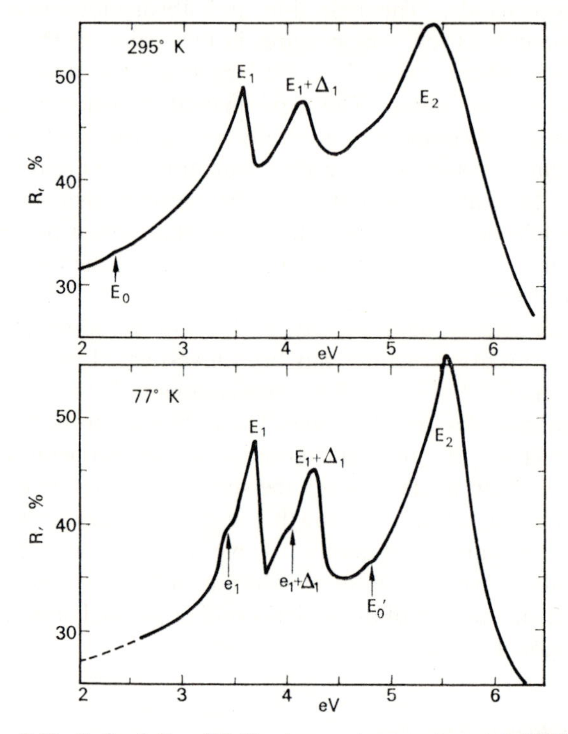

FIG. 3.12. Reflectivity of ZnTe at room temperature and 77°K.[56]

energy side of the $E_1 + \Delta_1$ peak designated by $e_1 + \Delta_1$ and with the same temperature dependence as $E_1 + \Delta_1$; it is thought the peak is associated with an $L_3 \rightarrow L_1$ transition at the zone edge. An e_1 peak also is partially resolved in the spectrum of ZnTe at 77°K and is illustrated in Fig. 3.12. Energy values have been obtained for the E_o, E_o', E_2, E_1', d_1 and d_2 transitions as well as those for E_1 and e_1. An $X_5 \rightarrow X_3$

(E_2) transition has been observed in HgTe and CdTe and values of temperature coefficient for several transitions have been quoted for all three tellurides. The d_1 and d_2 peaks in HgTe occur at considerably lower energy than in CdTe and ZnTe since the energy separation between the *d*-levels and the next *s*-levels is also lower in Hg (9 eV) than in Cd (12 eV) and Zn (11 eV).

Scouler and Wright[58] have studied the reflectivity of etched samples of HgSe and HgTe from 4–12 eV at room and liquid helium temperatures. The structure of the spectra is much more pronounced at liquid helium temperature and the spin–orbit splitting is observed at the X_5 valence band point in the $X_5 \rightarrow X_3$ and $X_5 \rightarrow X_1$ transitions. The $L_3 \rightarrow L_3$ transition with energy E_1 in their analysis seems more satisfactorily identified than by Cardona and Greenaway[56] whose spectral identification gives a spin-orbit splitting at the L_3 point of Δ_1 equal to the extremely large value of 1·25 eV. They observe strong peaks designated by E at 6·55 and 7·6 eV in HgTe and HgSe respectively, which they cannot identify although they suggest it may result from the band overlap which characterizes these semimetals.

Other reflectivity studies of the cubic II–VI crystals[57, 59, 60] have provided further confirmation of the interpretation of the early results and have made comparisons between compounds which can take both the cubic and hexagonal modifications. The study of the epitaxially deposited thin films of cubic CdS[59] from 2–22 eV was the first on the cubic modification of cadmium sulphide and produced reflectivity spectra of the same general form as other zinc blende materials. Table 3.6 combines the results of the different investigators for the zinc blende structured II–VI compounds at room and low temperatures and also compares them with those for Si and GaAs.

3.4.2. *Wurtzite Structure*

Cardona[62] was the first to record investigations of the reflection properties of ZnS, CdS, CdSe with the wurtzite structure at room temperature. The spectra observed, resembled the general form obtained for the zinc blende structures and peaks could be identified with the analogous energy band transitions in zinc blende.

Table 3.6 Energy Values in electron volts associated with the Peaks in the Reflectivity Spectrum of Silicon, Gallium Arsenide, and the II–VI Compounds with the Zinc Blende Structure.

These results were obtained at liquid nitrogen or helium temperature, except for those marked by an asterisk which are for room temperature measurements

Energy Designation \ Compound	Si[14]	GaAs[14]	HgTe	HgSe	CdTe	CdS	ZnTe	ZnSe	ZnS
E_0	3·6	1·55	− 0·14[20]	− 0·15[20]	1·59[56]	2·50*[59]	2·37[56]	2·8[55], 2·7*[31]	3·94[8]
E_0'	3·43	4·2	4·10[56]	5·20[8]	5·20[56]	6·20*[59]	4·82[56]	7·6[8]	5·8*[9]
E_1	3·7	2·99	2·21[56]	2·82*[55]	3·44[56]	5·49*[59]	3·71[56]	4·85[56], 4·75*[31]	5·3*[9]
e_1		2·6	2·00[56]		3·19[56]		3·48[56]		
E_1'	5·5	6·6	7·5[58]	8·3[58]	7·0[61], 6·76*[56]	9·18*[59]	7·30[61], 6·90*[56]	9·1*[31]	9·5*[8], 9·8*[9]
E_2	4·4	5·12	5·0[58]	5·7[58]	5·49[56]	7·95*[59]	5·54[56]	6·4*[31]	7·0, 7·4*[9]
E_2			5·4[58]	6·45[58]	5·99[56]				
d_1			9·55[58]	9·7[58]	10·1*[59]	11·9*[59]	10·6*[56]	10·6*[57]	10·8*[9]
d_2			11·1[58]	11·1[58]	13·8*[56]	14·4*[59]	14·6*[56]	13·8*[57], 13·5*[31]*	13·8*[9]
F_2						9·7*[59]			7·9*[9]
Δ_0	0·044	0·35	1·0[60]	0·5[60]	0·81[56]	0·06[60]	0·91*[56]	0·43[56], 0·45*[31]	0·6[8]
Δ_1	0·030	0·24	0·64*[55]	0·3[58]	0·58[56]	0·04[60]	0·58*[56]	0·35*[31]	0·05[51]
			0·75[58]						
$-dE_1/dT \times 10^4$		4·2[55]	4·5[55]	4·6[55]	5·5[56]		6·0[56]	5·2[55]	
$-dE_2/dT \times 10^4$			5·1[56]		4·1[56]		6·0[56]		

Subsequent investigations have used polarized radiation in order to detect effects which result from the anisotropic nature of the wurtzite structure.[57, 61, 63] Cardona[8] also notes the result of very limited measurements on zinc oxide in which the Γ_{15}–Γ_1 (E_o) transition requires an energy of 3·44 eV, the Γ_{15}–Γ_{15} transition an energy of 5·3 eV and the magnitude of the spin-orbit splitting is 0·0087 eV. An interesting study of epitaxial thin films of cubic and hexagonal CdS has permitted a direct comparison of the energy peaks in the respective reflection spectra.[59] The most notable feature in the comparison is that the E_1 peak in wurtzite is split into two components, A and B, which result from the non-equivalence of the wurtzite and zinc blende Brillouin zones. Figure 3.13 compares the reflectivity spectra of cubic and hexagonal films of CdS at room temperature. Other features of the wurtzite spectra are that the F_1 shoulder appears on the lower energy side of the E_2 peak and the F_3 shoulder on the high energy side of the E_1 peak, while the F_2 shoulder observed on the high energy side of the E_2 peak in zinc blende disappears. Cardona and Harbeke[9] observed a splitting in the A component of the E_1 peak, the magnitude of which was 0·23 eV roughly $\frac{2}{3}$ of Δ_o, the spin-orbit splitting for $\mathbf{k} = 0$. Table 3.7, after Cardona and Harbeke[9, 59], compares the energies of the peaks for ZnS, CdS, CdSe determined from the reflectivity and also includes the Kramers–Kronig derivation of the absorption index from reflectivity at room temperature.[64]

3.5. Band Structure

It has been seen in the preceding sections that the nature of energy transitions at different points in **k**-space is reasonably well identified in most of the II–VI compounds. However, the knowledge of band structure away from the principal symmetry points still tends to be more of a qualitative than a quantitative nature. The first principles orthogonalized plane waves (OPW) calculation of band structure has so far only been applied to the diamond structured elemental semiconductors diamond, silicon, germanium and grey tin by Herman.[65, 66] More recently Herman[67] has added an empirical crystal potential correction based on firmly established experimental data to correct for

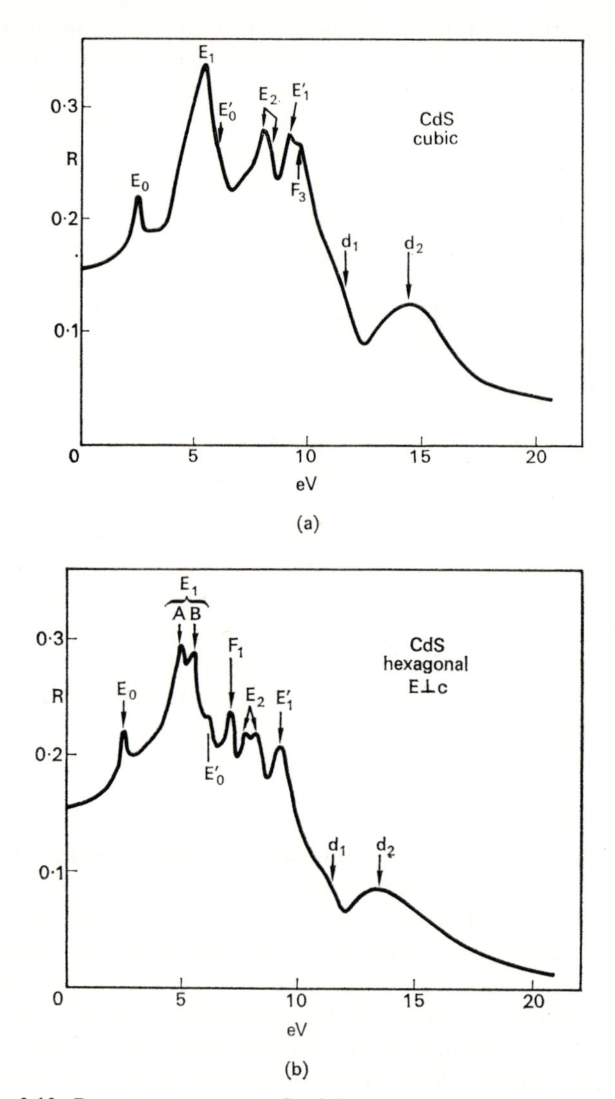

FIG. 3.13. Room temperature reflectivity spectra of cubic and hexagonal epitaxially grown CdS thin films.[59]

the free electron exchange approximations in his first principles calculations. The studies using the modified OPW method have been restricted to the group IV homopolar semiconductors although Herman has proposed an extension of the work to the zinc blende and wurtzite structures. It should be noted that the OPW method suffers two disadvantages for heavier elements or compounds; (1) the computational difficulties increase as the number of core levels grow and (2) anisotropy effects arise with the increase in core states.

The majority of the knowledge of band structure in the II–VI compounds has been derived from the semi-empirical pseudopotential method.[68,69] In the pseudopotential method an approximated Schrodinger equation with a potential term in it is used. The secular equation which results depends only on the pseudopotential form factors which are derived from the optical data available. Clearly inaccurate interpretation of optical results can lead to discrepancies in the pseudopotential derivation of band structure. The studies of Cohen and Bergstresser have included II–VI compounds with the zinc blende and wurtzite structures. Figure 3.14 (a), (b), (c) illustrate the band structure determined by Cohen and Bergstresser[68] for CdTe, ZnSe and ZnTe respectively, which take the zinc blende structure. Figure 3.15 (a) and (b) show similarly determined band structures for CdSe and ZnS which take the wurtzite structure.[69] It should be noted that in these two figures no account has been taken of spin-orbit splitting effects. Other studies of the band structure in II–VI compounds include two cellular methods which depend on non-overlapping charges for ZnS[70, 71] and the self-consistent OPW method for CdS.[72] Rossler and Lietz[70] in their calculations for ZnS locate the valence band maximum at Λ, a result which is contrary to experimentally observed data and is possibly explained by incorrect choice of atomic radii. Further, they disagree with Cardona and Harbeke's[5] interpretation of the 5·8 eV peak which they attribute to an L_1–L_3 transition and they associate the 8·4 eV peak with the Γ_{15}–Γ_{15} transition. Eckelt[71] has obtained results for ZnS which show good agreement with those of Cohen and Bergstresser,[68] a comparison of which is shown in Fig. 3.16.

The energy band structures of HgTe and HgSe have not been represented with absolute certainty, although developments have fallen

TABLE 3.7. PEAK ENERGIES FROM ABSORPTION INDEX (k) AND REFLECTIVITY (R) DATA AT ROOM TEMPERATURE IN ZnS, CdS AND CdSe (IN eV)[9,59]

		ZnS			CdS			CdSe	
			Hexagonal			Hexagonal		Hexagonal	
		Cubic	$\mathbf{E}\parallel c$	$\mathbf{E}\perp c$	Cubic	$\mathbf{E}\parallel c$	$\mathbf{E}\perp c$	$\mathbf{E}\parallel c$	$\mathbf{E}\perp c$
E_0	k	3·66	3·74	3·78	2·55				
	R	3·76	3·88	3·87	2·50	2·53	2·53	1·9	1·85
E_0'	k	5·82	5·86	5·80	6·4	6·2	6·1	6	
	R	5·79	5·76	5·74	6·2	6·3	6·2	6	
E_1 A	k	5·3					4·98		4·31, 4·08
	R	5·4		5·5			4·93		4·30, 4·07
B	k				5·3	5·48	5·50	4·98	4·85
	R		5·6		5·49	5·52	5·50	5·00	4·82
E_1'	k	9·65	9·56	9·43	8·8	9·27	9·00	8·75	8·50
	R	9·78	9·73	9·61	9·18	9·35	9·15	8·63	8·35
E_2	k	7·03, 7·35	7·01, 7·50	7·00, 7·5	7·5	7·8	7·8	7·30	7·54
	R	6·99, 7·41	6·98, 7·56	7·00, 7·52	7·95	8·04, 8·35	8·00, 8·35	7·50	7·55
d_1	k	10·6			11·6	11·4			
	R	10·8	10·8	10·8	11·9	11·5		10·75	10·6
d_2	k	14·3			13·3				
	R	13·8	13·8	13·8	14·4	14·0	14·0	14·0	14·0
F_1	k			6·6					6·9
	R			6·6			7·12		6·8
F_2	k	7·9			9·6				
	R	7·9			9·7				
F_3	R					9·8	9·8	9·35	9·2

broadly along the lines suggested by Harman *et al.*[21] and represented in Fig. 3.5. Groves[20, 73] slightly modified Harman's proposed structure to include the inversion of the Γ_1 conduction band state with the Γ_{15} valence band states; this modification leads to a negative energy gap

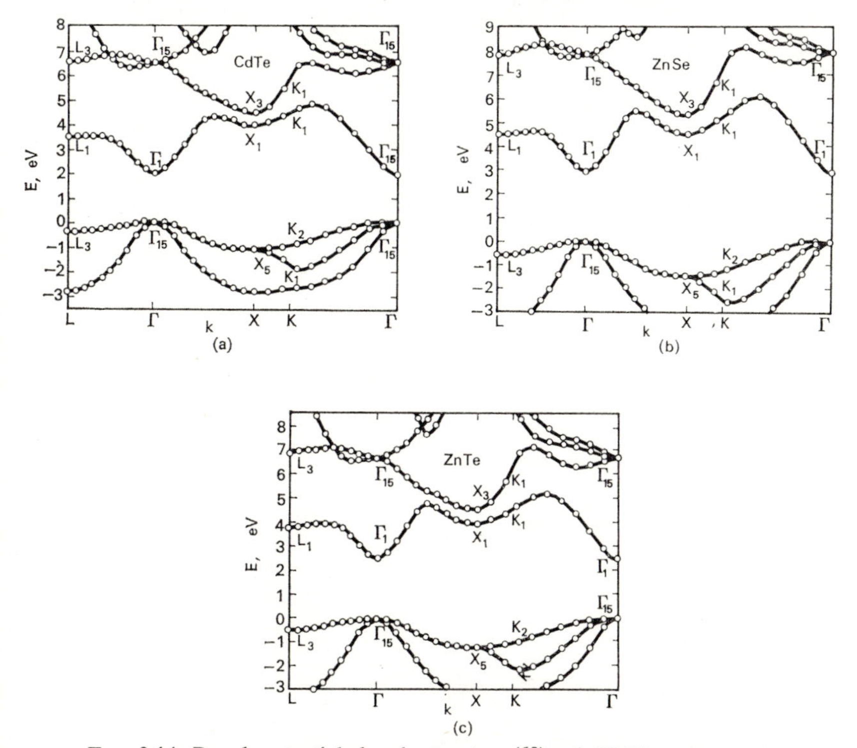

FIG. 3.14. Pseudopotential band structures[68] of II–VI compounds with zinc blende structure, (a) CdTe, (b) ZnSe, (c) ZnTe.

difference of approximately $-0{\cdot}15$ eV for the two compounds. Such a result does not conflict with the experimental observations of Harman[21] on the HgTe–CdTe alloys in which, if it is assumed that the energy gap varies linearly with composition, a negative E_G would be obtained for HgTe.

It would seem unfortunate to leave this section on band structure without mention of effective mass. Effective mass measurements have been made on most of the II–VI compounds by one method or another.

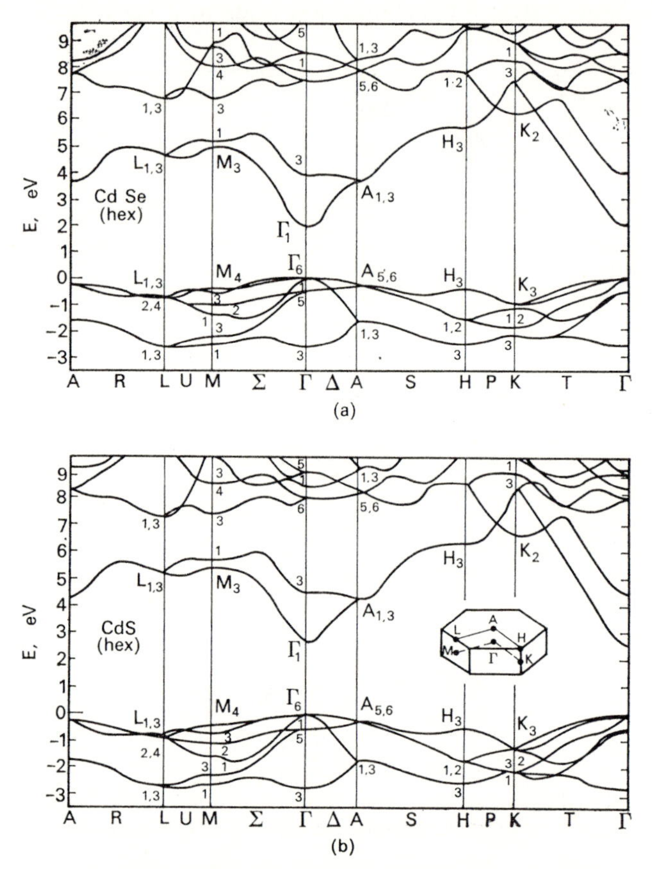

FIG. 3.15. Pseudopotential band structure of II–VI compounds with the wurtzite structure, (a) CdSe, (b) CdS.[69]

The effective mass is used to provide information about the curvature of the bands in which carriers give rise to conduction and hence clarify the detail of the band structure at certain symmetry points. The methods employed to determine effective mass include both optical and electrical techniques, which it is not proposed to consider in this chapter. However, for reference purposes the effective mass values of the II–VI compounds are included here in Table 3.8.

TABLE 3.8. EFFECTIVE MASS OF CARRIERS IN II–VI COMPOUNDS IN UNITS OF THE FREE ELECTRON REST MASS[14]

	Exciton reduced mass	Electron effective mass m_n^*	Hole effective mass, m_p^*	
			Light	Heavy
CdTe	0·1	0·096, 0·11[j]	0·30	
CdSe	0·13	0·13	0·45	1
CdS	0·18	0·153–0·171	0·7	5
ZnTe			0·15–0·31[g]	2[g]
ZnSe	0·13[i]	0·17[h]–0·15[f]	0·6	
ZnS	0·18	0·27	0·58	
ZnO	0·31	0·38	1·8	
HgTe		0·052[e], 0·03[c]–0·18[d]	0·16[a]	
HgSe		0·007[a]–0·068[b]	0·17[a]	

(a) Harman, T. C., and Strauss, A. J., *J. Appl. Phys.* **32,** 2265 (1961).
(b) Whitsett, C. R., *Phys. Rev.* **138A,** 829–39 (1965).
(c) Stradling, R. A., *Proc. Phys. Soc.* **90,** 175–80 (1967).
(d) Szymanska, W., *Phys. Stat. Sol.* **23,** 69–73 (1967).
(e) Szymanska, W., Sniadower, L., and Giriat, W., *Phys. Stat. Sol.* **10,** K11–13 (1965).
(f) Aven, M., and Segall, B., *Phys. Rev.* **130,** 81–91 (1963).
(g) Watanabe, N., *J. Phys. Soc. Japan* **231,** 713–24 (1966).
(h) Marple, D. T. F., *J. Appl. Phys.* **35,** 1879–82 (1964).
(i) Riccius, H. D., and Turner, R., *J. Phys. Chem. Solids* **28,** 1623–4 (1967).
(j) Marple, D. T. F., *Phys. Rev.* **129,** 2466–70 (1963).

Mercury telluride and mercury selenide exhibit strong interaction between conduction and valence bands and this leads to a very small effective mass at the bottom of the conduction band. Hence as the carrier concentration increases so the effective mass increases and it is not difficult to observe the mass change by a factor of ten. The two light hole effective masses quoted for ZnTe are for the two curvatures at the Γ_{15} symmetry point.

3.6. Dielectric Constant and Refractive Index

It is necessary to give at least a brief mention to dielectric constant and refractive index in order to complete a chapter on fundamental optical properties.

Both the electronic and ionic polarization contribute to the dielectric behaviour of the II–VI compounds. The increasing magnitude of the degree of ionicity in these compounds relative to the group IV

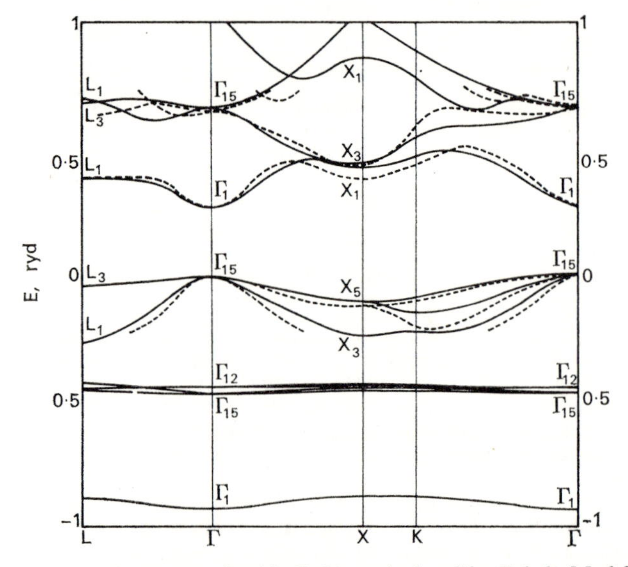

FIG. 3.16. Band structure of cubic ZnS as calculated by Eckelt, Madelung and Treusch[71] and the results of Cohen and Bergstresser[72] (dashed lines).

elements and III–V compounds leads to significant differences between the static and optical (high frequency) dielectric constants. The static dielectric constant ε_s shows an appreciable temperature dependence at room temperature which is associated with transverse optical phonon interaction, whereas as at low temperatures ε_s is approximately constant. The decrease in ionic character along a series ZnO, ZnS, ZnSe, ZnTe seems to be borne out by the decrease in $\varepsilon_s - \varepsilon_\infty$ values in this sequence. The anisotropic nature of the compounds that take the wurtzite structure has been observed in both the static dielectric constant and the optical dielectric constant as derived from refractive index measurements.[14, 74] The behaviour of the refractive index on the long wavelength side of

TABLE 3.9. DIELECTRIC CONSTANTS AND REFRACTIVE INDICES FOR II–VI COMPOUNDS

Compound	Structure	Dielectric constant ε_s	Dielectric constant ε_∞	Refractive index
CdS	W	$\varepsilon\|_c$ 8·64[6] $\varepsilon\perp_c$ 8·28[6]	5·24[14]	2·30[14] ($\lambda = 2\ \mu$m)
CdSe	W	$\varepsilon\|_c$ 9·25[6] $\varepsilon\perp_c$ 8·75[6]	6·4[74]	2·55[74] ($\lambda = 0{\cdot}86\ \mu$m)
CdTe	ZB	9·65[6]	7·13[14]	2·67[14]
ZnO	W	$\varepsilon\|_c$ 8·5[6] $\varepsilon\perp_c$ 8·15[6]	4·59[14]	2·14[14]
ZnS	ZB	8·1[14]	5·13[14]	2·26[14] ($\lambda > 2\ \mu$m)
ZnSe	ZB	8·66[6]	5·90[75]	2·43[75]
ZnTe	ZB	9·67[6]	7·28[75]	2·68[75]

$\varepsilon_{\|c}$ and $\varepsilon_{\perp c}$ are the static dielectric constants for propagation parallel and perpendicular to the *c*-axis in the wurtzite structure respectively.

the absorption edge has been studied by several authors.[74–77] Table 3.9 compares the static and optical dielectric constants for the II–VI compounds. Figure 3.17 illustrates dispersion curves for CdSe, CdTe, ZnSe and ZnTe in the region of the absorption edge.[74–76] The dispersion curves for these compounds all take a similar form with the refractive index increasing as more electrons are excited into the conducting state.

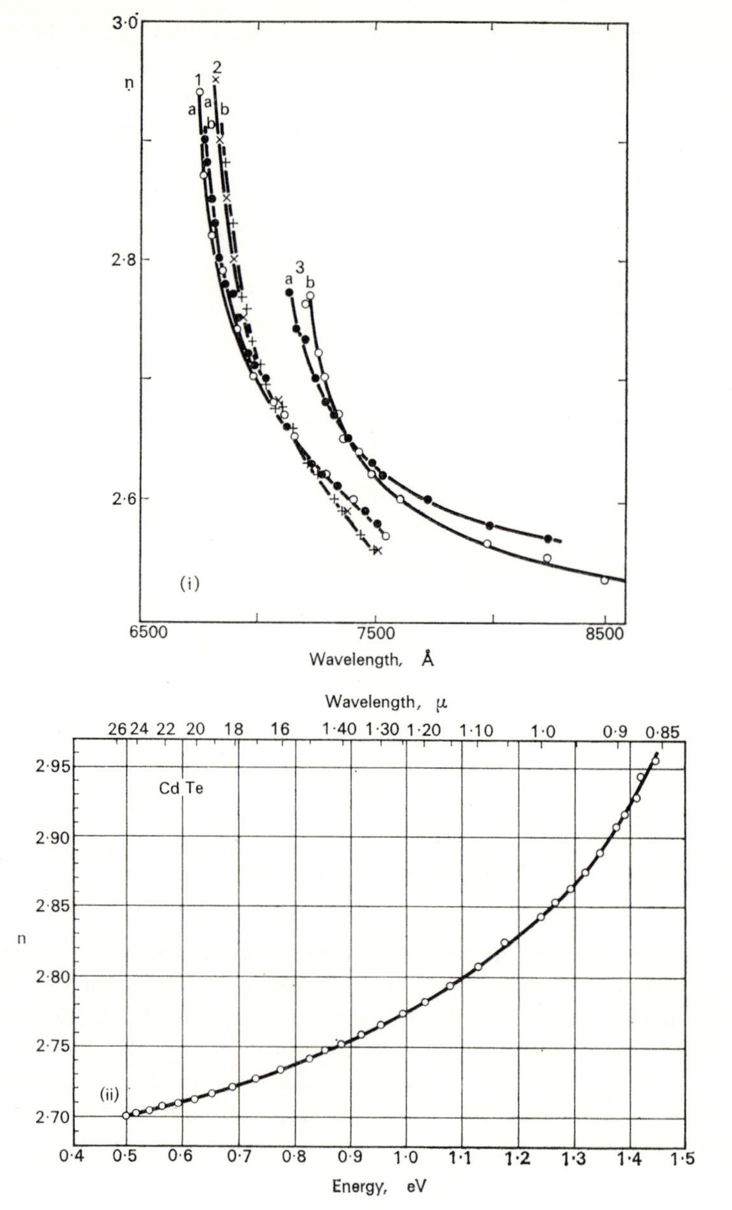
n
1
2
a
a
b
b
3
a
b
3·0
2·8
2·6
(i)
6500
7500
8500
Wavelength, Å
Wavelength, μ
26 24 22 20 18 16 1·40 1·30 1·20 1·10 1·0 0·9 0·85
Cd Te
2·95
2·90
2·85
2·80
2·75
2·70
n
(ii)
0·4 0·5 0·6 0·7 0·8 0·9 1·0 1·1 1·2 1·3 1·4 1·5
Energy, eV

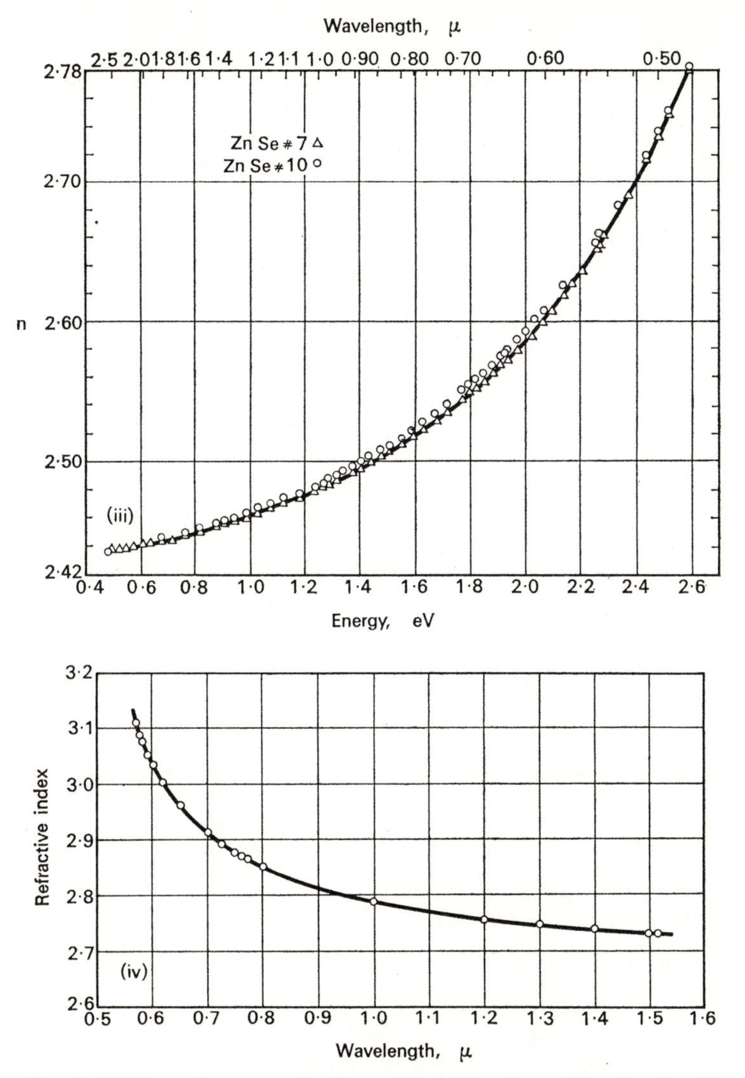

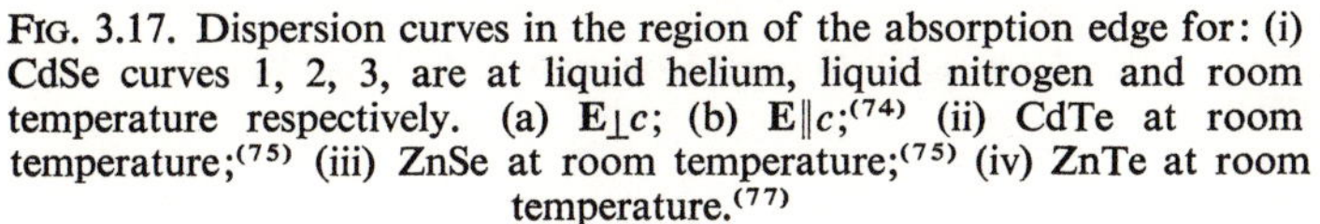

FIG. 3.17. Dispersion curves in the region of the absorption edge for: (i) CdSe curves 1, 2, 3, are at liquid helium, liquid nitrogen and room temperature respectively. (a) $\mathbf{E} \perp c$; (b) $\mathbf{E} \parallel c$;[74] (ii) CdTe at room temperature;[75] (iii) ZnSe at room temperature;[75] (iv) ZnTe at room temperature.[77]

References

1. Moss, T. S., *Optical Properties of Semiconductors*, pp. 1–14, 83–90. Butterworths, London, 1961.
2. Lax, B., and Mavroides, J. C., *Applied Optics* **6,** 647–59 (1967).
3. Cardona, M., Pollak, F. H., and Shaklee, K. L., *J. Phys. Soc. Japan* **S21,** 89–93 (1966).
4. Gutsche, E., and Lange, H., *Phys. Stat. Sol.* **22,** 229–33 (1967).
5. Lange, H., and Henrion, W., *Phys. Stat. Sol.* **23,** K67–70 (1967).
6. Segall, B., and Marple, D. T. F., *Physics and Chemistry of II–VI Compounds*, pp. 319–81 (M. Aven and J. S. Prener eds.). North Holland, Amsterdam, 1967.
7. Birman, J. L., Samelson, H., and Lempicki, A., *G.T. and E. Journal* **1,** 2–15 (1961).
8. Cardona, M., *J. Phys. Chem. Solids* **24,** 1543–55 (1963).
9. Cardona, M. and Harbeke, G., *Phys. Rev.* **137A,** 1467–76 (1965).
10. Edwards, A. L., and Drickamer, H. G., *Phys. Rev.* **122,** 1149–57 (1961).
11. Langer, D., *Proc. 7th Int. Conf. Physics Semiconductors*, pp. 241–4. Dunod, Paris, 1964.
12. Thomas, D. G., *J. Phys. Chem. Solids* **15,** 86–96 (1960).
13. Reynolds, D. C., Litton, C. W., and Collins, T. C., *Phys. Stat. Sol.* **9,** 645–84 (1965).
14. Reynolds, D. C., Litton, C. W., and Collin, T. C., *Phys. Stat. Sol.* **12,** 3–55 (1965).
15. Edwards, A. L., Slykehouse, T. E., and Drickamer, H. G., *J. Phys. Chem. Solids* **11,** 140–8 (1959).
16. Knell, R. L., and Langer, D. W., *Phys. Letters* **21,** 370–1 (1966).
17. Madelung, O., *Physics of III–V Compounds*, pp. 26–41. Wiley, New York, 1965.
18. Reynolds, D. C., *et al.*, *J. Phys. Soc. Japan* **S21,** 143–7 (1966).
19. Groves, S., and Paul, W., *Phys. Rev. Letters* **11,** 194–5 (1963).
20. Groves, S., and Paul, W., *Proc. 7th Int. Conf. Physics Semiconductors*, pp. 41–49. Dunod, Paris, 1964.
21. Harman, T. C., *et al.*, *Solid State Comm.* **2,** 305–8 (1964).
22. Wright, D. A., *Brit. J. Appl. Phys.* **16,** 939–42 (1965).
23. Wang, S., *Solid State Electronics*, pp. 118–22. McGraw Hill, New York, 1966.
24. Mitra, S. S., *Phys. Rev.* **132,** 986–91 (1963).
25. Mitra, S. S., *J. Phys. Soc. Japan* **S21,** 61–66 (1966).
26. Balkanski, M., and Besson, J., *J. Appl. Phys.* **32,** 2292–7 (1961).
27. Spitzer, W. G., *J. Appl. Phys.* **34,** 792–5 (1963).
28. Deutsch, T., *J. Appl. Phys.* **33,** 751–2 (1962).
29. Marshall, R., and Mitra, S. S., *Phys. Rev.* **132,** 563 (1963).
30. Collins, R. J., *J. Appl. Phys.* **30,** 1135–40 (1959).
31. Aven, M., Marple, D. T. F., and Segall, B., *J. Appl. Phys. Suppl.* **32,** 2261–5 (1961).
32. Dietz, R. E., Thomas, D. G., and Hopfield, J. J., *Phys. Rev. Letters* **8,** 391–3 (1962).

33. JOHNSON, F. A., *Progress in Semiconductors* **9,** 181–235. Heywood, London, 1965.
34. HALSTED, R. E., *Physics and Chemistry of II–VI Compounds*, pp. 385–431 (M. Aven and J. S. Prener eds.). North Holland, Amsterdam, 1967.
35. BALKANSKI, M., *Proc. 7th Int. Conf. Phys. Semiconductors*, pp. 1021–36. Dunod, Paris, 1964.
36. KLICK, C. C., *J. Opt. Soc. Amer.* **41,** 816–23 (1951).
37. PEDROTTI, L. S., and REYNOLDS, D. C., *Phys. Rev.* **120,** 1664–9 (1960).
38. COLBOW, K., *Phys. Rev.* **141,** 742–9 (1966).
39. MAEDA, K., *J. Phys. Chem. Solids* **26,** 1419–30 (1965).
40. VAN DOORN, C. Z., *Philips Res. Repts.* **21,** 163–79 (1966).
41. HANDELMAN, E. T., and THOMAS, D. G., *J. Phys. Chem. Solids* **26,** 1261–8 (1965).
42. REYNOLDS, D. C., PEDROTTI, L. S., and LARSON, O. W., *J. Appl. Phys. Suppl.* **32,** 2250–4 (1961).
43. PEDROTTI, F. L., and REYNOLDS, D. C., *Phys. Rev.* **127,** 1584–6 (1962).
44. LANGER, D. W., PARK, Y. S., and EUWANA, R. N., *Phys. Rev.* **152,** 788–96 (1966).
45. SAMELSON, H., and LEMPICKI, A., *Phys. Rev.* **125,** 901–8 (1962).
46. HALSTED, R. E., and SEGALL, B., *Phys. Rev. Letters* **10,** 392–5 (1963).
47. DUTTON, D., *Phys. Rev.* **112,** 785–92 (1958).
48. COLLINS, R. J., and HOPFIELD, J. J., *Phys. Rev.* **120,** 840–2 (1960).
49. ANDRESS, B., and MOLLWO, E., *Naturwiss,* **46,** 623–4 (1959).
50. THOMAS, D. G., and HOPFIELD, J. J., *Phys. Rev.* **128,** 2135–48 (1962).
51. REYNOLDS, D. C., and LITTON, C. W., *Bull. Am. Phys. Soc.* **9,** 224 (1964).
52. REYNOLDS, D. C., LITTON, C. W., and WHEELER, R. G., *Proc. 7th Int. Conf. Phys. Semiconductors*, pp. 739–744. Dunod, Paris, 1964.
53. COLLINS, T. C., LITTON, C. W., and REYNOLDS, D. C. *Proc. 7th Int. Conf. Physics Semiconductors*, pp. 745–9. Dunod, Paris, 1964.
54. HALSTED, R. E., and AVEN, M., *Phys. Rev. Letters* **14,** 64–65 (1965).
55. CARDONA, M., and HARBEKE, G., *J. Appl. Phys.* **34,** 813–18 (1963).
56. CARDONA, M., and GREENAWAY, D. L., *Phys. Rev.* **131,** 98–103 (1963).
57. BALKANSKI, M., and PETROFF, Y., *Proc. 7th Int. Conf. Physics Semiconductors*, pp. 244–50. Dunod, Paris, 1964.
58. SCOULER, W. J., and WRIGHT, G. B., *Phys. Rev.* **133,** 736–9 (1964).
59. CARDONA, M., WEINSTEIN, M., and WOLFF, G. A., *Phys. Rev.* **140,** 633–7 (1965).
60. SOBOLEV, V. V., *Optics and Spectroscopy* **18,** 456–9 (1965).
61. BALKANSKI, M., *J. de Physiques* **28,** C3, 36–42 (1967).
62. CARDONA, M., *Phys. Rev.* **129,** 1068–9 (1963).
63. CARDONA, M., *Solid State Comm.* **1,** 109–15 (1963).
64. CARDONA, M., and GREENAWAY, D. L., *Phys. Rev.* **133A,** 1685–97 (1964).
65. HERMAN, F., *Proc. I.R.E.* **43,** 1703–32 (1955).
66. HERMAN, F., *Rev. Mod. Phys.* **30,** 102–21 (1958).
67. HERMAN, F., *et al., J. Phys. Soc. Japan* **S21,** 7–14 (1966).
68. COHEN, M. L., and BERGSTRESSER, T. K., *Phys. Rev.* **141,** 789–96 (1966).
69. BERGSTRESSER, T. K., and COHEN, M. L., *Phys. Rev.* **164,** 1069–80 (1967).
70. ROSSLER, U., and LIETZ, M., *Phys. Stat. Sol.* **17,** 597–604 (1966).
71. ECKELT, P., MADELUNG, O., and TREUSCH, J., *Phys. Rev. Letters* **18,** 656–8 (1967).

72. Collins, T. C., Euwena, R. N., and DeWitt, J. S., *J. Phys. Soc. Japan* **S21,** 15–19 (1966).
73. Groves, S. H., Brown, R. N., and Pidgeon, C. R., *Phys. Rev.* **161,** 779–93 (1967).
74. Parsons, R. B., Wardzynski, W., and Yoffe, A. D., *Proc. Roy. Soc.* **262A,** 120–31 (1961).
75. Marple, D. T. F., *J. Appl. Phys.* **35,** 539–42 (1964).
76. Rambauske, W. R., *J. Appl. Phys.* **35,** 2958–9 (1964).
77. Sliker, T. R., and Jost, J. M., *J. Opt. Soc. Amer.* **56,** 130–1 (1966).

CHAPTER 4

LUMINESCENCE

4.1. General Features of Luminescence

The word luminescence describes the emission of radiation which results when a material adjusts itself from an excited to a ground state. The material must previously have been stimulated into the excited state. The form of stimulation is denoted by a prefix to the word luminescence and the phenomena of particular relevance to II–VI compounds are photoluminescence, thermoluminescence and electroluminescence. Cathodo-luminescence, which might be classed as a subdivision of electroluminescence since an electric field is required to drive the electrons from the cathode to the luminescent anode, has been studied with reference to the commercial requirements of fast response and decay times.[1] Limited studies of mechanically excited luminescence (triboluminescence) in CdS and ZnS have also been made as instanced by the work of Warschauer and Reynolds,[2] and Obrikat.[3] Luminescent emission is somewhat arbitrarily divided into fast and slow processes, fluorescence referring to emission which occurs within a microsecond of excitation and phosphorescence to time delays of greater than a microsecond. Needless to say great flexibility is exercised in the use of these definitions and hence what might be thought of as abuse particularly to the word fluorescence frequently occurs. The fact that most luminescent materials exhibit phosphorescence has led to the name "phosphors" being applied to them generally.

The II–VI phosphors have a terminology used for them which differs from that of the conventional semiconductor. Two types of impurities are introduced into II–VI compounds to modify their luminescent behaviour and they are called activators (acceptors) and coactivators (donors). The common activator elements are copper, silver and manganese, which occupy group II cation lattice sites and behave as deep acceptor levels. The activator impurity has traditionally been thought of as the site at which the recombination of electron and hole occurs with radiative emission of energy and is often called the luminescent centre. Typical coactivator elements are the halides (Cl, Br, I) occupying anion sites and aluminium, gallium and indium occupying cation sites and these form deep donor levels below the conduction band edge. The depth of these levels is such that thermal activation at room temperature is not sufficient to completely remove the electrons from the levels which have consequently become known as electron traps. In addition to defects created by deliberately added impurities there always exist in crystallites natural defects. These natural defects are usually of zinc and sulphur vacancies and they provide self activation and coactivation for the luminescent II–VI compounds. The zinc and sulphur vacancies behave as double acceptor and double donor impurities respectively. The assumption that sulphur vacancies provide the self-coactivation is not without ambiguity since zinc or cadmium interstitials would exhibit similar donor-like characteristics, although perhaps on atomic size grounds they are rather large to provide a simple interstitial state. Sulphur interstitials are very much more unlikely, since sulphur has a large atomic radius and is rarely obtained in monatomic form. The emitting II–VI compounds have the tendency that if they are doped with one impurity type only, then the charge on the resultant defect is compensated by the self generation of the appropriate natural defect with charge of the opposite sign.

The location of the activator and coactivator levels in the forbidden energy gap and the energy transitions which involve luminescent emission can be represented basically in terms of three general models seen in Fig. 4.1. The Schön–Klasens model regards luminescence as a result of radiative recombination of an electron from the conduction band with a localized acceptor level above the valence band.[4, 5] Lambe and Klick[6] proposed a model which describes the luminescent transition in terms of a free hole combining with a trapped electron at a level below the conduction band. The Lambe–Klick model stemmed from difficulties in the explanation of a longer decay time for the photoconductivity than for luminescence in CdS on a Schön–

Klasens model. The Prener–Williams model was established to account for the absence of luminescence in ZnS with copper at random substitutional sites as obtained by the transformation of Zn^{65} to copper under neutron irradiation.[7] They considered that a localized association of activator and coactivator was necessary to give rise to luminescent emission.

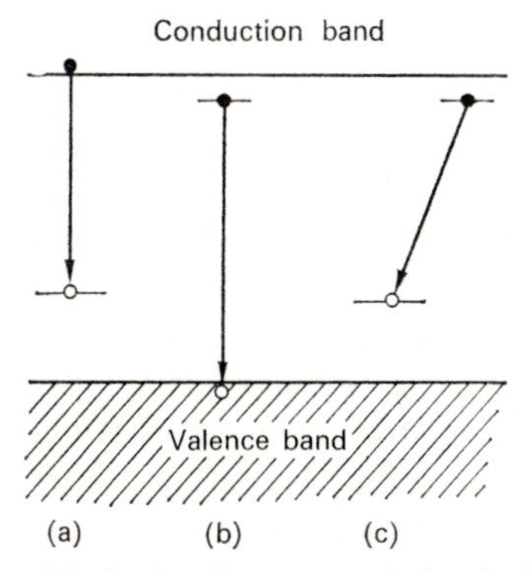

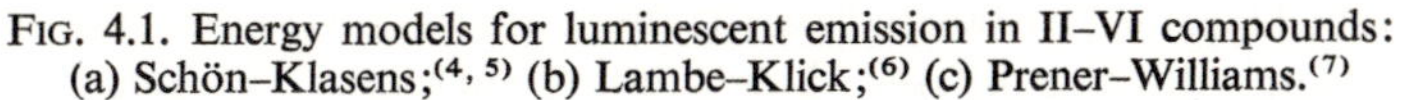
FIG. 4.1. Energy models for luminescent emission in II–VI compounds: (a) Schön–Klasens;[4, 5] (b) Lambe–Klick;[6] (c) Prener–Williams.[7]

A broad basis has been provided above for a discussion of the particular attributes of the II–VI phosphors. The discussion will be divided into three sections; photoluminescence (mainly fluorescence), thermoluminescence and electroluminescence. A limitation has been made to three types of luminescence since they represent the bulk of the luminescent studies on the II–VI compounds and are particularly pertinent to technological developments.

4.2. Photoluminescence

The section on photoluminescence will look at the different means of activation in ZnS-type phosphors and try to find a theoretical model which satisfies the observed experimental behaviour.

4.2.1. *Self-activated Emission*

Fluorescence emission has been observed in several II–VI compounds which have not been deliberately activated. The emission peak shifts slightly with respect to wavelength or energy for changes in the coactivator impurity type as is illustrated for ZnS in which the peak wavelengths range from 4560 to 4710 Å at 80°K.[8] Similar behaviour has been found in CdS,[9, 10] ZnSe,[11] ZnTe,[10] CdTe.[10]

The mechanism of self-activated luminescence in II–VI compounds generally, however, is uncertain, although in ZnS consistent interpretation has been achieved on the basis of a single model. An association between neighbouring lattice sites of the doubly ionized zinc vacancy acceptor defect and an ionized impurity donor defect (Cl_S^+ or Al_{Zn}^+) has been suggested.[12] The resultant singly ionized acceptor associated complex permits fluorescent emission to be considered on a Schön–Klasens model. The emission peak shift to longer wavelengths with a change in coactivator from anion (Cl_S) to cation (Al_{Zn}) substitution may also be explained on an associated centre model. Since chlorine (Cl_S) and a zinc vacancy (V_{Zn}) can take up adjacent lattice sites and an (Al_{Zn}–V_{Zn}) associated pair cannot, then the two acceptor levels that result will have different hole binding energies. The hole binding energy will be smaller the closer the charged coactivator to the zinc vacancy and therefore the closer will the acceptor energy level be to the valence band. Thus a Schön–Klasens model in conjunction with an associated centre explains the observed emission peak shift with coactivator change. A similar analysis has also been applied to ZnSe with the same conclusions.[11] The polarized nature of the radiation emitted from chlorine coactivated ZnS, observed by Koda and Shionoya[13] in their study of single crystal material has tended to confirm a localized centre of the kind (Cl_S–V_{Zn}). Numerous investigations of the electron paramagnetic resonant behaviour of photoexcited compounds has lent further support to the associated centre model. The shift of the self-activated emission to shorter wavelengths with increasing temperature bears some similarity to that observed in the edge emission of CdS in section 3.3.1 and may be explained by the break up of some of the associated centres. Recent investigations of the radioactive decay of

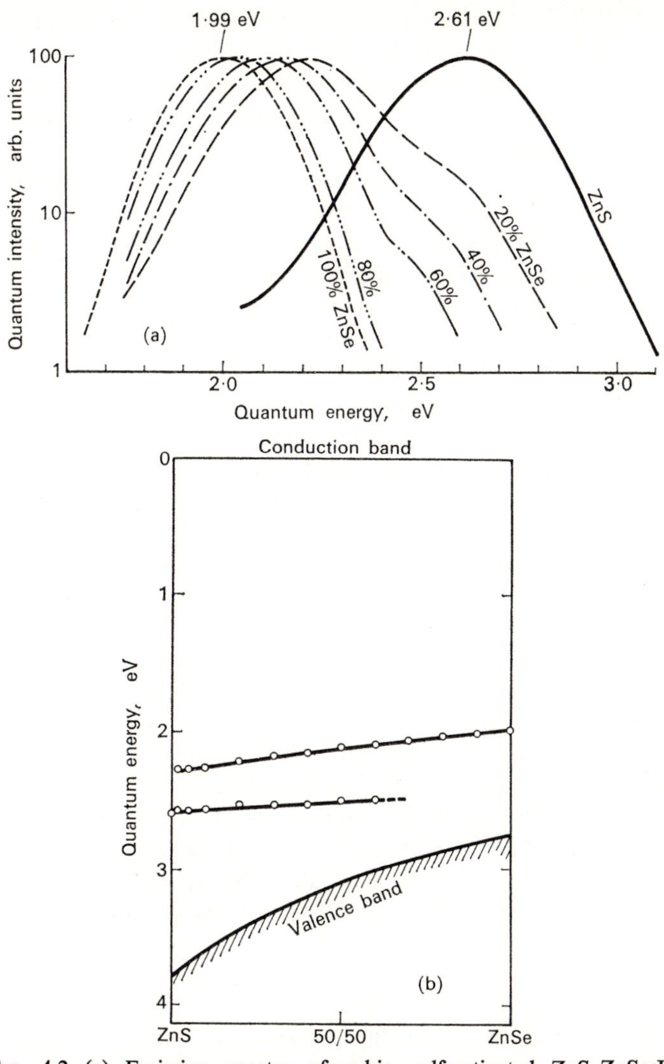

FIG. 4.2. (a) Emission spectra of cubic, self-activated ZnS–ZnSe:Br phosphors excited by ultraviolet radiation ($\lambda = 365$ mμm) at 77°K. (b) Recombination scheme of self-activated ZnS–ZnSe: Br phosphors at 77°K.[15]

S^{35} to Cl^{35} in ZnS crystals have shown a massive increase in the blue self-activated emission as the half life of the S^{35} (87 days is approached.[14] Such a result adds further evidence for an associated pair (V_{Zn}–Cl_S) to provide the recombination level at which the self-activated emission occurs.

Extensive studies of the emission spectra at low temperatures of self-activated ZnS and alloys formed between the II–VI compounds have indicated the presence of a higher energy emission peak which requires band gap energy exciting radiation for its stimulation.[8, 10, 14] In ZnS it has been suggested that the peak results from a transition of the Lambe–Klick type in which a valence band hole recombines with an electron trapped at a donor level. The emission energy appeared to depend only on the location of the coactivator level as determined from thermoluminescence measurements (section 4.3). This high energy emission has disappeared at room temperature when presumably thermal effects have reduced the probability and incidence of the transition.

The general conclusions that may be drawn about the behaviour of self-activated ZnS, ZnSe, CdS and CdSe phosphors with halide coactivation is that the emission is of the broad band type when excited by 365 mμm radiation at liquid nitrogen temperature. The gradual replacement of zinc by cadmium in either sulphide or selenide alloys results in a steady shift of the emission band to lower energies. The replacement of sulphur by selenium in ZnS creates little shift in the emission band although gradual replacement of this band by another occurs; similar behaviour is observed in the CdS–CdSe and ZnSe–ZnTe alloy systems.[15] Figure 4.2 illustrates, (a) the change from one emission peak to the other in ZnS–ZnSe:Br phosphors at 77°K; (b) the recombination scheme in the same alloy system.

4.2.2. *Copper-activated Emission*

Copper activation is probably the most used form of sensitization in efficient visible emitting phosphors. The green emission at room temperature from ZnS in particular is the most familiar characteristic of copper activation. An interesting feature of copper activation is that when it is added in excess of its solubility in ZnS it influences the crystal structure taken; the mechanism is thought to be that excess copper precipitates as cubic copper sulphide which forms a nucleation centre for the growth of cubic zinc sulphide at temperatures (> 1200°C) where the hexagonal modification is the stable form in the pure compound.(16) The precipitation of excess copper has gained prominence in the explanation of the electroluminescence mechanism in ZnS (section 4.4).

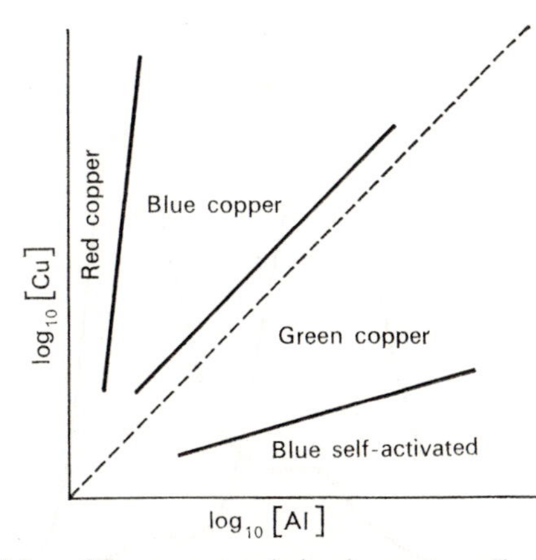

FIG. 4.3. Division of fluorescent emission in copper-activated, aluminium-coactivated ZnS.(8)

Van Gool(8) has investigated the fluorescent emission of copper-activated ZnS with both halide and aluminium coactivation in great detail. He summarized these investigations with an excellent generalized diagram seen for ZnS: Cu, Al in Fig. 4.3. It seems logical in view of the depth of observation on ZnS to analyse the different coloured emissions in copper-activated ZnS and then to make comparisons with other copper-activated II–VI compounds.

(a) *Copper green and blue emission.* These two emissions are sensitive to small changes in copper concentration, the presence of excess copper relative to the added coactivator concentration enhances the dominance of the blue peak. This is instanced in Fig. 4.4, where a change in activator concentrations from 10^{-4} to $1{\cdot}5 \times 10^{-4}$ Cu with aluminium concentration constant at 10^{-4} causes a shift from green peak to blue peak domination.[17, 18] Phosphors coactivated with aluminium have a weaker blue emission than those coactivated with chlorine which might be explained by the possibility that the solubility of chlorine and aluminium in ZnS are dependent on the copper concentration. The

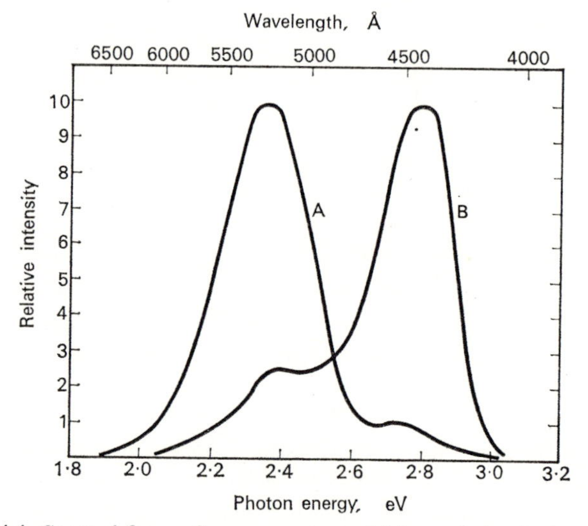

FIG. 4.4. Spectral form of copper green and blue emission in hexagonal ZnS at 80°K.[18] A: ZnS:10^{-4} Cu, 10^{-4} Al. B:ZnS:$1{\cdot}5 \times 10^{-4}$ Cu, 10^{-4} Al.

temperature dependence of both the blue and green emission peaks is to larger wavelengths with increasing temperature, i.e. in the same direction as the energy gap.

Lehmann[19] has made detailed studies of the copper blue and green emission in ZnS and several alloy systems formed between ZnS, CdS, ZnSe, CdSe and ZnTe. The emission from the compounds ZnS, ZnSe, CdS and CdTe consists of two broad peaks separated by about 0·4 eV. In the alloys, replacement of zinc by cadmium in the sulphides or the selenides resulted in a steady shift of both emission peaks to lower energies, whereas replacement of sulphur by selenium or selenium by tellurium caused very little change in the emission peak energy.

The mechanism of the copper blue and green emissions has not been resolved with any great certainty. Curie[20] has suggested that the green emission results from a transition between a coactivator level and an associated copper activator level which is of the Prener–Williams type (section 4.1). The copper blue emission seems to be generally ascribed to a transition between an electron at the bottom of the conduction band and a copper level, Schön–Klasens type transition. The infrared quenching of both the blue and green emissions at the same wavelength observed by Browne[21] was thought to lend support to the above, although it could be argued that the association of copper with a coactivator should shift the copper level relative to the top of the valence band. Time resolved spectral observations of the copper green emission in ZnS indicate that the decay of luminescence shows an appreciable peak energy shift to lower energy after a time delay of a millisecond.[22] Such behaviour is interpretable on an association model in which the pairs separated by greater distances take longer to decay and have smaller transition energies.

Both the copper blue and green emissions from hexagonal zinc sulphide single crystals have been observed to be preferentially polarized perpendicular to the *c*-axis.[23, 24] Birman[25] proposed an explanation for the polarization of blue emission that required the emitting centre to have the same symmetry as the energy bands of the hexagonal zinc sulphide. Such a proposal may equally well be applied to the green emission. The requirement is relatively straightforward for the blue emission based on a Schön–Klasens model and the copper level has to take the same symmetry Γ_9 as the uppermost valence band. However, for the green emission with a Prener–Williams model to describe it the coactivator level must take the symmetry of the lowest conduction band (Γ_7) and again the copper level requires the same symmetry as the uppermost valence band (Γ_9). An alternative model for the green emission has been suggested with the copper centre taking the same symmetries as the three valence bands. Thus the blue and green emissions would result from different symmetry states of the copper centre.

Lehmann[19], as a consequence of his studies on the blue and green emission in II–VI alloys, has proposed a qualitative explanation of his results which assumes a predominantly ionic bond type. He associated the cation with the conduction band and the anion with the valence band. The luminescence is ascribed to electron transitions from the conduction band into previously emptied levels between the activator ion and the four surrounding chalcogenide anions. Replacement of zinc by cadmium causes only a variation of the conduction

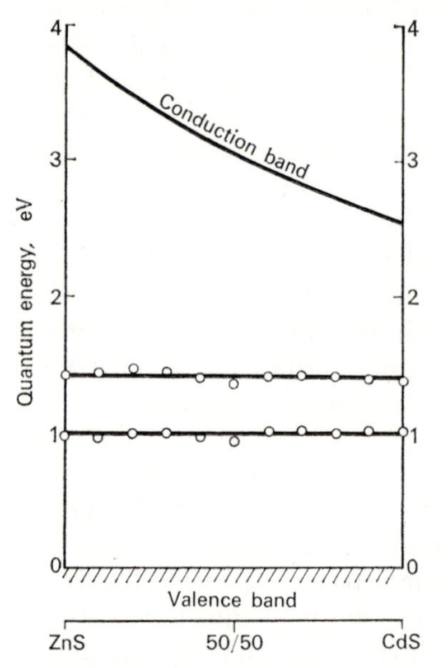

FIG. 4.5. Energy scheme for copper levels giving rise to copper green and blue emission in hexagonal (Zn, Cd) S: Cu, Cl at 77°K.[19]

band while the activator electron levels and the valence band remain unchanged. Hence the energy separation of the activator level and the valence band is independent of the Zn/Cd ratio, Fig. 4.5. Replacement of sulphur by selenium only affects the energy separation between the activator level and valence band. These observations are in broad agreement with the earlier investigations on the ZnS–ZnSe:Cu system.[10, 25, 26]

(b) *Copper red emission.* Detailed studies of the copper red emission in ZnS indicate that it only appears when a hydrogen sulphide atmosphere is used during preparation in order to remove chlorine and oxygen impurities.[8, 28] The emission peaks in hexagonal ZnS occur at 6740 and 6960 Å for temperatures of 300 and 80°K respectively. Additional evidence that this copper emission occurs in other II–VI compounds is the observation of Halsted *et al.*[10] on ZnSe of a peak at 9900 Å at 25°K. The dependence of the copper red emission peak energy on composition in the ZnS–ZnSe system follows a similar pattern to that of the copper green and blue emission. The principal requirements for copper red emission are a large excess of copper relative to added coactivators, Fig. 4.3, and the absence of halide coactivator (aluminium coactivator has little effect on the emission). The polarization of the emitted radiation suggests that a dipolar effect along the symmetry axis of the centre causes the luminescence and the behaviour of the copper red emission is thus similar to the self-activated emission with chlorine coactivation.

The centre responsible for the emission is thought to be a copper atom associated with a sulphur vacancy (Cu_{Zn}–V_S). Dieleman[29] has concluded as a result of electron spin resonance studies of the copper red emission in ZnS:Cu that the associated centre behaves as a singly ionizable donor defect (Lambe–Klick model). The emission may be described as self-coactivated and is complementary to the self-activated emission described on the basis of an associated centre behaving as a singly ionizable acceptor (Schön–Klasens model). The temperature shift of the emission given above would be explained satisfactorily on this model.

(c) *Copper infrared emission.* Both copper doped ZnS and CdS exhibit emission bands in the infrared in addition to those in the visible.[21, 30] The preparation conditions that lead to strong infrared emission are excess sulphur pressures and it is thought that this reduces the concentration of sulphur vacancies available to associate with copper which may then exist in the uncompensated state. Isolated copper behaves as an acceptor and gives rise to infrared emission when an electron is captured—the emission is often called acceptor emission.

The infrared emission in ZnS is divided into two significant peaks (0·84 and 0·75 eV) and a small peak at 0·69 eV. Figure 4.6 illustrates the emission in ZnS at 4°K.[31] Infrared emitting phosphors can be subdivided into two categories on the basis of the stimulation required. The first category are excited by infrared radiation peaking at 1·63 and 0·92 eV and they contain copper in excess of coactivators. The second category requires simultaneous stimulation with infrared and ultraviolet radiation and they have similar concentrations of activator and coactivator; coexistence of the infrared emission with visible emission occurs under these circumstances.

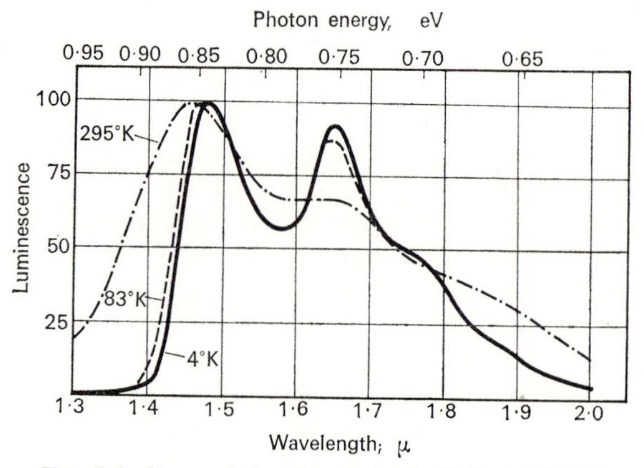

FIG. 4.6. Copper infrared emission in ZnS at 4°K.[31]

Thermoelectric studies on ZnS crystals which exhibit infrared emission under infrared excitation indicate that they are *p*-type semiconducting[32] and hence justify the use of the term acceptor emission. Radioactive decay of Zn^{65} in ZnS to give stable Cu^{65} produced a linear growth in acceptor emission with copper concentration over the range 10^{-7} to 10^{-5} g copper/g ZnS.[33] The evidence is seen to point fairly conclusively to a copper acceptor level as the source of the infrared emission. Several approaches have been made to the origin of the two main peaks and include (1) the ground and first excited state of a hole bound to a substitutional copper impurity, (2) the different ionized states of the copper atom on an ionic model, (3) a single copper level and transitions to the different valence bands.[18] The latter approach by Bryant and Cox[34] permits the three peaks to be described by transitions from a single copper level to crystal field and spin-orbit split valence bands at the Γ point. The agreement between the infrared peak energy separations and $\Delta_{so} = 0{\cdot}092$ eV and $\Delta_{cr} = 0{\cdot}055$ eV taken from Table 3.5 for hexagonal zinc sulphide is good.

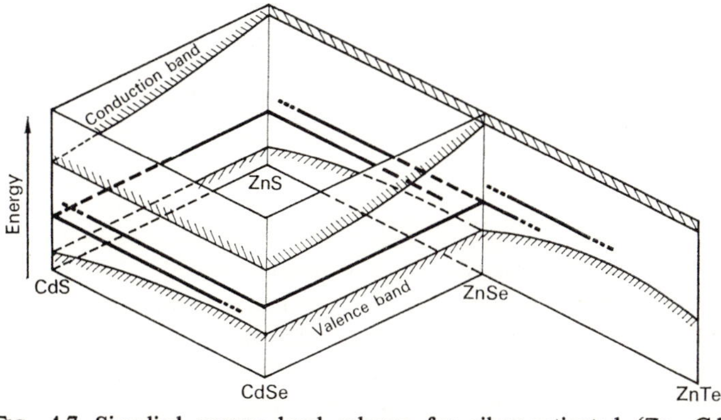

FIG. 4.7. Simplied energy level scheme for silver-activated (Zn, Cd) (S, Se, Te) phosphors.[35]

(d) *Analogous emission.* Silver and gold belong to the same group in the periodic table as copper and as such might be expected to have similar emission characteristics with perhaps one or two exceptions. ZnS:Ag exhibits a blue emission with a peak energy of 2·8 eV at 77°K which is virtually independent of both the coactivator concentration and the crystal structure.[35] The behaviour of the silver blue band is almost identical to that of the copper blue emission and the mechanism is therefore likely to be the same (Schön–Klasens transition). A weak green emission has been observed with ZnS:Ag but it is not clear whether this results from copper impurity or a silver-coactivator associated level. The blue visible emission from the silver-activated

II–VI compounds, ZnS, ZnSe, CdS and CdSe, consists essentially of one peak and the behaviour of this peak in alloys formed between these compounds and also ZnTe is illustrated in Fig. 4.7.[35] A striking feature about silver-activated alloys in which substitution on the anion site occurs is that after approximately a 10% addition of the heavier anion the equivalent of the silver green emission peak has at least the same intensity as the blue emission. Gold-activated CdS exhibits the equivalent of copper green and blue emission plus a third emission shifted to the red which occurs when a trivalent coactivator balances the gold activator.[36] Infrared luminescence of the kind observed in copper-activated ZnS and CdS has not been observed in silver or gold doped sulphides or selenides. The self-coactivated copper red emission observed in ZnS has been found in silver- and gold-activated ZnS shifted to the yellow and orange spectral regions respectively.[29]

4.2.3. *Transition Metal Associated Emission*

The general characteristic of a transition metal is its incomplete *d* electron shell which gives rise to a particular form of emission spectrum when such an element is incorporated into a crystalline host lattice.

Manganese is the most investigated transition metal activator in ZnS-type lattices. The manganese Mn^{2+} ion occupies a substitutional anion lattice site and in ZnS gives rise to orange luminescence peaking at 2·12 eV (858 mμm) with a half width of 0·23 eV. The absorption spectrum for Mn^{2+} in ZnS consists of five peaks in the visible region at 2·34, 2·45, 2·60, 2·87 and 3·12 eV or in terms of wavelengths at 530, 506, 477, 432 and 398 mμm.[37, 38] The absorption peaks result from *d–d* transitions normally forbidden in the free ion state and consequently are of low intensity with relatively long decay constants $\sim$ 2 ms at 77°K. The lowest energy absorption peak (2·34 eV) is barely resolved except close to liquid helium temperatures.

Both McClure[38] and Langer and Ibuki[39] have observed fine structure in the emission and absorption spectra of manganese-activated ZnS at 4·2°K. Figure 4.8 illustrates the fine structure on the short wavelength edge of the emission peak for ZnS:Mn; there is a periodicity in the fine structure and Langer and Ibuki[39] have related this to the absorption of phonons of energy 0·037 eV. McClure[38], on the other hand, suggested that the fine structure resulted from Mn–Mn ion pairs. A recent study of the build-up and decay times of the excitation spectra associated with ZnS doped with varying manganese concentrations did not show any detectable variation in times, thus supporting the phonon absorption idea.[40] However, with large manganese concentrations ($> 3 \times 10^{-2}$ g Mn/g ZnS) two additional emission bands do appear with maxima at 635 and 745 mμm and they predominate over the orange emission in the 50–250°K temperature range. The excitation spectra for all three emission bands are the same.

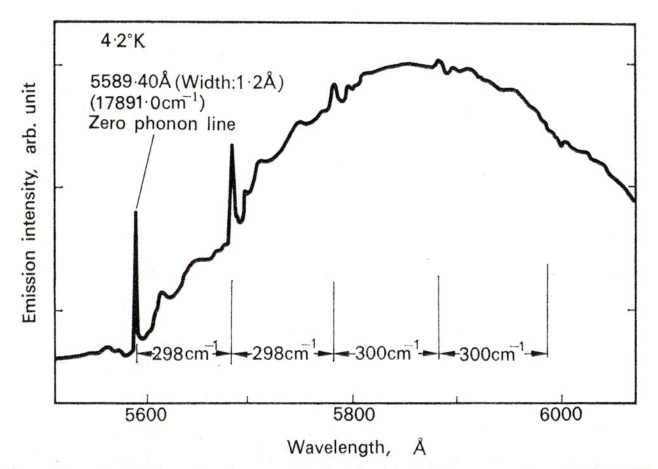

FIG. 4.8. Emission spectrum of a cubic ZnS:Mn crystal at 4·2°K.[39]

A full interpretation of both the excitation and emission spectra is based on the electron states of the free ion Mn^{2+} with $3d^5$ configuration and the interaction between these states and the crystalline field of the host lattice. A brief indication only of the interpretation will be recorded here and the reader is referred to more detailed accounts.[18, 41] The simplest analysis is given in terms of the absorption spectra where the peaks for increasing values of energy originate from transitions between the ^6S ground state and lowest excited states ^4G, ^4P, ^4D, ^4F, etc.

Splitting of these states occurs in the constraining crystalline field and leads to a more complex spectrum. Excitation from the ground state to states associated with the $3d^4 4s$ configuration require energies of 7 eV or more and thus were not observed in the spectral region investigated.

Avinor and Meyer[42, 43] have looked in detail at both the emission and excitation spectra of vanadium in CdS, ZnS, CdSe and ZnSe. A single emission peak at approximately 0·6 eV (2 μm) is observed in all four compounds when vanadium activated. Addition of copper and silver raises the efficiency of the emission and it is suggested the vanadium exists in the trivalent V^{3+} ($3d^2$) state. The vanadium activator kills the emission that normally appears from copper and silver which merely act as sensitizers for the vanadium emission. Table 4.1 lists the emission peak positions in electron volts for the vanadium activated CdS, ZnS, CdSe and ZnSe under different sensitized conditions.[42] Absorption peaks characteristic of the vanadium are found at 1·1 and 1·6 eV in the excitation spectra independent of the sensitizing conditions. The quantum efficiencies of emission in vanadium-activated zinc and cadmium sulphides have been investigated with different exciting energies 3·40 eV (366 mμm), 2·85 eV (436 mμm), and 2·27 eV (546 mμm).[44] The highest efficiency of 49% was obtained for a silver-sensitized phosphor, ZnS:5×10^{-5} Ag, when excited by 4·30 eV radiation. Allen[45] has interpreted both the emission and excitation spectrum in terms of the crystal field interaction with the V^{3+} ion along lines similar to that applied to manganese activation.

Other transition elements which have been used quite often for activation in ZnS type phosphors are Cr^{2+}, Fe^{2+}, Co^{2+}, Ni^{2+}.[18, 46] Iron-activated ZnS has recently been studied at 5°K and infrared emission in the range 0·29 to 0·38 eV has been observed in addition to the usual red emission which peaks at 1·88 eV (660 mμm).[47] The infrared emission which is characteristic of the Fe^{2+} ion in the crystalline lattice shows fine structure indicative of phonon absorption from the ZnS lattice. Slack and O'Meara[47] also noted similar infrared emission in ZnS containing traces of cobalt. Sensitization of fluorescent emission in transition metal-activated phosphors can be effected by the addition of copper or silver which generally have their own characteristic emission quenched.

4.2.4. *Emission Activated by Other Impurity Elements*

Extremely interesting observations have been made of the fluorescent emission of ZnS and CdS doped with tellurium.[48] Tellurium is a neutral impurity which therefore does not require charge compensation and differs sufficiently in its electronegativity value from that of the

TABLE 4.1. EMISSION PEAK POSITIONS IN eV FOR VANADIUM-ACTIVATED PHOSPHORS

Activators	Temperature °K	CdS	ZnS	CdSe	ZnSe
V	300	0·59	0·66	0·54	~ 0·51 0·63
	80	0·60	0·68	0·56	~ 0·51 0·64
V, Ag	300	0·55	~ 0·57 0·63	0·51	~ 0·46 0·62
	80	0·57	0·60 ~ 0·68	0·53	~ 0·46 0·57 ~ 0·65
V, Cu	300	0·57	0·59	0·52	~ 0·63
	80	~ 0·56 0·59	0·62	0·54	~ 0·50 0·60
V, Au	300	0·59			
	80	0·55 ~ 0·60			

sulphur to occupy an energy state in the forbidden energy gap of the sulphide phosphor. Orange fluorescence with a peak at 2·04 eV is observed in CdS:10^{-4} Te phosphors and the emission varies in the same manner as the energy gap with composition for $Zn_xCd_{1-x}S$ phosphors for x in the range of 0 to 0·4.

The origin of the emission is uncertain although a Lambe–Klick model has been suggested to describe it. Alkali metal doped phosphors indicate that Li and Na function as acceptors in ZnS, CdS, ZnSe;[49] potassium has proven acceptor behaviour in ZnSe only. The joint presence of iron and an alkali metal in a II–VI phosphor results in the appearance of the Fe^{3+} ion. These observations are in agreement with similar results obtained for lithium doped zinc oxide.[50]

Luminescence resulting from rare-earth ions in II–VI compounds has received little investigation because of the difficulties of incorporation of rare-earth ions into ZnS-type lattices. The luminescent spectra for rare-earth doped compounds are thought to be due to f–f transitions of the ions and are made of numerous narrow lines in the visible and infrared spectral regions. Ibuki and Langer[51] have studied the emission from $ZnS:Tm^{3+}$ and $ZnS:Ho^{3+}$ and have been able to identify the transition from the ^1G excited state to the ^3H ground state with the most intense line (478 mμm) in $ZnS:Tm^{3+}$. Anderson[52] has discussed the green luminescence of $ZnS:Tb^{3+}$ in terms of carrier transport without a definite conclusion. Sensitization of rare-earth-activated emission has also been observed in ZnS and $Zn_xCd_{1-x}S$ doped with copper or silver[53] and in ZnSe doped with copper or lithium.[54]

4.3. Thermoluminescence

The phenomenon of thermoluminescence (thermal glow) is used to gain information about the electron trapping states in phosphorescent materials. To activate thermoluminescent emission the following sequence of events is necessary; the phosphor is thermally cleaned by heating to the maximum temperature to be used, cooled to the lowest temperature (usually the boiling point of liquid nitrogen), irradiated for several minutes with radiation of energy greater than the phosphor's energy gap, held at 80°K while the phosphorescent emission decays and finally heated rapidly and at a constant rate to the maximum temperature. The luminescent emission occurs during the rapid heating and bursts of light are emitted over several temperature ranges to indicate non-thermal equilibrium trap emptying. By detailed study of the form of the emission peaks, the reaction kinetics and energy depths of the electron traps below the conduction band can be deduced. Thermoluminescence emission is difficult to observe in materials with appreciable conductivities and in such examples the analogous technique of thermally stimulated currents is used. The theory behind thermoluminescence is well established, although the unambiguous interpretation of experimental results is rarely achieved. A brief account of the theory of thermoluminescence and the method of electron trap depth evaluation will be given in the succeeding subsection followed by the results obtained for different impurities in the II–VI compounds.

4.3.1. *Theoretical Foundations of Thermoluminescent Emission*

In phosphorescent materials the net rate of release of electrons from traps is taken to be equal to the luminescent emission. It is assumed that radiative recombination occurs at a single luminescent centre as the dominant process for electrons which are not retrapped. Dependent upon the variable of the electron release rate one observes phosphorescence (time) or thermoluminescence (temperature), thus both phosphorescence and thermoluminescence provide similar information about the electron trapping states; a detailed discussion of this subject as a whole is given in references.[55–57] The theory of emission will be considered for three conditions which represent one intermediate and two extremal cases and these are: (a) monomolecular kinetics, (b) bimolecular kinetics, and (c) fast retrapping processes. It will be assumed for the purpose of analysis that the trapping states result from one type of defect and therefore have a single energy value.

(a) *Monomolecular kinetics.* The electrons are taken to be thermally excited from the traps and proceed via the conduction band to the radiative recombination centres, hence the luminescent emission intensity I is equal to the rate of emission of electrons from the traps, $-dn_t/dt$, where n_t is the number of trapped electrons

$$I = -\frac{dn_t}{dt} = \frac{n_t}{\tau_t} \tag{4.1}$$

where τ_t is the lifetime of electrons in the traps

$$\frac{1}{\tau_t} = s\exp\left(-\frac{E}{kT}\right) \tag{4.2}$$

where $E = E_c - E_t$ is the depth of the trap level below the bottom of the conduction band and s is an escape probability factor equal to $N_c v S_T$, N_c is the effective density of states in the conduction band, v is the electron thermal velocity and S_T the electron capture cross-section of the trapping state.

At this point the analyses for phosphorescent and thermoluminescent emission diverge with the application of a constant heating rate to the cooled phosphor

$$\beta = \frac{dT}{dt}$$

Hence

$$\frac{dn_t}{n_t} = -\frac{s}{\beta}\exp\left(-\frac{E}{kT}\right)dT$$

and integration between the initial and final temperatures T_o and T, leads to

$$n_t = n_{t_o}\exp\left\{-\int_{T_o}^{T}\frac{s}{\beta}\exp\left(-\frac{E}{kT}\right)dT\right\} \tag{4.3}$$

where n_{t_o} is the number of electrons trapped at temperature T_o and the luminescent emission intensity is given by

$$I = n_{t_o}s\exp\left(-\frac{E}{kT}\right)\exp\left\{-\int_{T_o}^{T}\frac{s}{\beta}\exp\left(-\frac{E}{kT}\right)dT\right\} \tag{4.4}$$

Physically equation (4.4) implies that (1) at low temperatures the emission intensity rises exponentially as $\exp(-E/kT)$ with the second exponential approximately equal to unity, (2) at intermediate temperatures dependent upon the value of E the second exponential starts to decrease in value and causes the emission intensity to reach a maximum and (3) at high temperatures the second exponential decreases very rapidly and the intensity falls to zero. The unbroken curve in Fig. 4.9 illustrates the general form of a glow curve for monomolecular kinetics.

The temperature T_m of the maximum in the thermoluminescent emission (glow peak) is given by

$$\left(\frac{dI}{dT}\right)_{T_m} = 0$$

hence

$$E = kT_m \log_e \left(\frac{skT_m^2}{\beta E}\right) \tag{4.5}$$

An approximate estimate of E for inclusion in the logarithmic term permits E to be calculated provided that s is known.

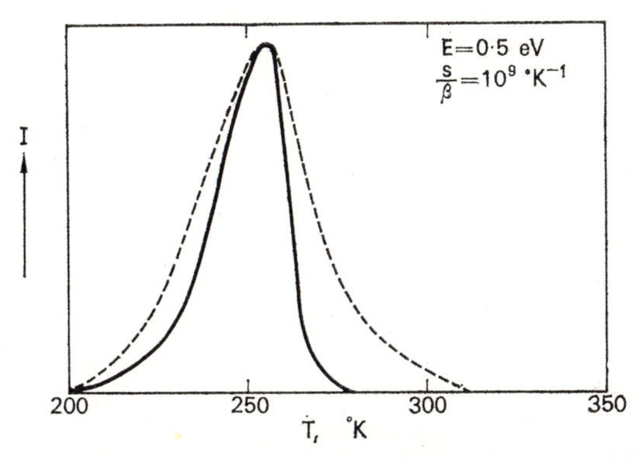

FIG. 4.9. Thermal glow curves for monomolecular (——) and bimolecular (----) kinetics with identical values of E, s and β.

(b) *Bimolecular kinetics.* Bimolecular kinetics for the purposes of this text refer to the condition that there are equal recombination and retrapping probabilities and also equal numbers of luminescent centres and electron traps. Thus, if the phosphor contains N_t electron traps of which n_t are filled by electrons, then there are $N_t - n_t$ empty traps and n_t luminescent centres. The probability that an electron will recombine with an empty luminescent centre is given by $n_t/N_t - n_t + n_t = n_t/N_t$ and the emission intensity is then

$$I = -\frac{dn_t}{dt} = \left(\frac{n_t^2}{N_t}\right) s \exp\left(-\frac{E}{kT}\right) \tag{4.6}$$

This leads to the following expression for the emission intensity

$$I = \frac{n_{t_o}^2}{N_t} s \exp\left(-\frac{E}{kT}\right)\left[1+\frac{n_{t_o}}{N_t}\int_{T_o}^{T}\frac{s}{\beta}\exp\left(-\frac{E}{kT}\right)dT\right]^{-2} \tag{4.7}$$

and the trap depth E as derived from the glow maximum condition is given by

$$E = kT_m \log_e\left(\frac{n_{t_o} s\, kT_m^2}{N_t \beta E}\right) \tag{4.8}$$

an effective escape probability factor $s_{\text{eff}} = n_{t_o} s/N_t$ replaces s in the expression derived for monomolecular kinetics, equation (4.5). It is clear from equation (4.8) that the extent to which traps are filled initially determines the glow maximum temperature T_m.

(c) *Fast retrapping.* Fast retrapping refers to the condition that electrons released from traps have a much higher probability of returning to the traps than of radiative recombination at a luminescent centre. This situation can almost be approximated to that of thermal equilibrium between electrons in the conduction band and those in trapping states and is therefore handled more satisfactorily in conductivity measurements as a function of temperature (thermally stimulated currents). Under thermal equilibrium conditions the magnitude of the Fermi energy $E_{fn} = E_c - E_F$ is given by

$$E_{fn} = kT_m \log_e \frac{N_c}{n_c} \tag{4.9}$$

Bube[58] has assumed that since the glow curve is almost symmetrical about the maximum, i.e. traps are half emptied at the maximum, then E_{fn} can be approximated to the trap depth E.

Haering and Adams[59] have derived an expression for the trap depth which is similar in form to those obtained for monomolecular and bimolecular kinetics.

$$E = kT_m \log_e \frac{N_c kT_m^2}{N_t \beta \tau E} \tag{4.10}$$

where τ is the recombination lifetime of electrons in the conduction band.

With fast retrapping the thermal emptying of a trap of depth E occurs at a very much higher temperature than for monomolecular or bimolecular conditions and luminescence processes are usually greatly impaired by quenching effects at high temperatures, thus rendering the technique as unsuitable. Electrical conductivity as a function of temperatures will provide the details about trapping states in these circumstances.

4.3.2. *Evaluation of the Thermoluminescent Emission Curves (Glow Curves)*

Garlick and Gibson[60] proposed an evaluation of the trap depth from the initial exponential rise of the glow curve. The method assumes that the term

$$\int_{T_0}^{T} \exp\left(-\frac{E}{kT}\right) dT$$

is approximately zero for values of T considerably less than that for the glow peak and hence the emission is described by $I = \text{constant} \times \exp(-E/kT)$. This approach is applicable to both monomolecular and bimolecular kinetics and a plot of $\log_e I$ versus $1/kT$ has a slope equal in magnitude to the trap depth E. With the trap depth thus determined the escape probability factor can in turn be derived from the glow maximum condition either from equation (4.5) or (4.8). Haake[61] has analysed the method in detail and concluded that the temperature range over which measurable emission occurs is only of the order of 10–15°K, a factor which considerably limits the accuracy with which E can be determined.

Several interpretations have been made on the basis of the glow maximum condition and the temperature at the half peak points.[62–64] However, they are all subject to very severe temperature limitations and will not be considered here. The area beneath the glow curve in conjunction with the glow maximum temperature has been used by several investigators to derive the trap depth. Halperin and Braner[65], in their analysis which requires an isolated peak, have attempted to find fairly accurately the form of the kinetics from the shape of the glow curve.

Hoogenstraaten[56] proposed an analysis of glow curves based on the variation of the glow maximum temperature with heating rate which permitted both the trap depth E and the escape probability factor s to be determined. The method requires the re-arrangement of equation (4.5) or (4.8) to the following form

$$\log_e\left(\frac{\beta}{T_m^2}\right) = \log_e\left(\frac{ks}{E}\right) - \frac{E}{kT_m} \qquad (4.11)$$

The plot of $\log_e(\beta/T_m^2)$ versus $1/kT_m$ has a slope of magnitude E and the value of E may be used to calculate s or s_{eff} dependent upon the kinetics. The principal limitation of this analysis is experimental in that heating rate changes of several orders of magnitude are required to produce accurately definable changes in T_m ($\sim$ 50°K). It must be added also that the assumption that $s = N_c v S_T$ remains approximately constant, with the peak temperature changes observed, is not without question.

Apprehensiveness about the idea of single trap depths lead Hoogenstraaten to analyse decayed and built up glow curves. A decayed glow curve is obtained by heating the excited phosphor until part of the emission has occurred, followed by rapid cooling and then observation of the glow curve. This procedure is repeated with the temperature, at which the first heating is stopped, increased so that progressively smaller emission is observed in the decayed glow curves. Built-up glow curves are derived by the converse procedure in which the duration of initial excitation is increased from small periods ($\sim$ 5 sec) to saturation time ($\sim$ 10 min). Decayed and built-up glow curves with the same total light emission should be identical if a single trap depth with constant escape probability factor is assumed. With monomolecular kinetics the decayed and built-up glow peaks should be diminished images of the fully excited glow peak. If the kinetics are bimolecular the glow peak temperature will depend on the initial trap filling according to a modified form of equation (4.8),

$$\log_e n_{t_o} T_m^2 = \log_e\left(\frac{N_t \beta E}{sk}\right) + \frac{E}{kT_m} \qquad (4.12)$$

Equation (4.12) indicates that a decrease in n_{t_o} will cause an increase in the glow maximum temperature T_m.

Experimental observations of ZnS:Cu, Cl and also other types of phosphor (Zn_2SiO_4:Mn, As) have shown a shift of decayed glow peaks to higher temperatures with decreasing total light emission. Figure 4.10 illustrates the behaviour of both decayed and built-up glow curves for ZnS:5 × 10^{-5} Cu fired in an HCl atmosphere at 1000°C; n_{t_0} in equation (4.12) is directly proportional to C_{00} in the fully excited phosphor and to C_0 when the phosphor is only partially excited. The trap depth derived from a plot of $\log_e C_0 T_m^2$ versus $1/T_m$ for decayed glow

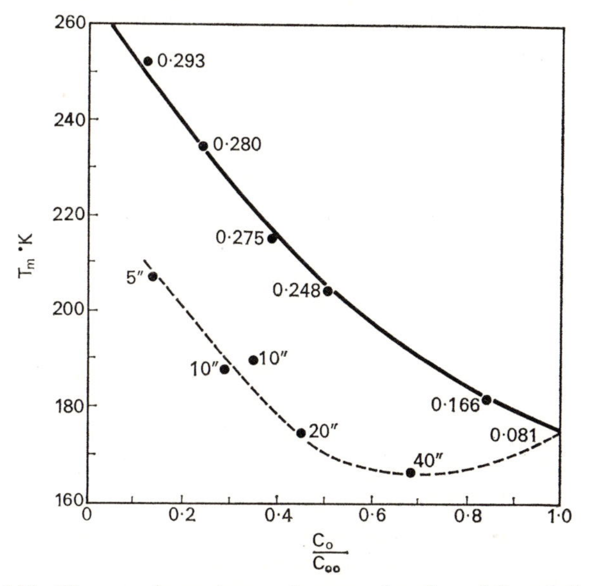

FIG. 4.10. Glow maximum temperature as a function of the relative light sum for decayed (——) and built-up (----) glow curves of a thin layer of ZnS:5 × 10^{-5} Cu (fired in HCl at 1000°C).[56]

curves of the ZnS phosphor was 0·045 eV which is rather small for a large energy gap semiconductor with an appreciable ionic character in its bond type. An energy state at 0·045 eV below the conduction band edge which is not appreciably emptied until 250°K suggests that thermal equilibrium conditions between the trap and the conduction band are almost in evidence, a characteristic of fast retrapping 4.3.1 (c). The behaviour of the built-up glow curve, however, does not conform with that of the decayed glow curve and indicates that it may not simply be a question of the kinetics which give rise to the observed temperature shift of the glow maximum. The built-up glow curves show a fall in glow maximum temperature with decreasing excitation times down to 40 sec followed by a rise in T_m with shorter excitation times. An example of the difference in glow maximum temperatures with the emitted relative light intensity, $C_o/C_{oo} = 0{\cdot}02$ is decayed $T_m = 239°K$ and built-up $T_m = 202°K$. The explanation for these differences requires different storage mechanisms for the two methods of analysis and it is quite possible that a distribution in both s and E values occurs.

A distribution in trap depths has been indicated by the values of E derived by Garlick and Gibson's method from the low temperature part of the decayed glow curves. These values are shown in Fig. 4.10 and E increases from 0·081 eV for the fully excited curve to 0·293 eV for the curve after decay of 12·2% of maximum emission. More recent confirmation for a range of trap depths in these materials has come from Bube's analysis[(66)] of decayed conductivity glow curves in CdS; the change in initial slope of $\log_e I$ versus $1/T$ obtained by Bube[(67)] is illustrated in Fig. 4.11. A distribution in traps seems probable therefore and it is extremely improbable that such a distribution in traps will have the same value of escape probability factor. In fact if the escape probability factor s was constant for all E values then this would cause T_m to shift more rapidly to higher temperatures with lower degrees of filling. The built-up curves show a decrease in T_m with decreased filling of traps except for very short excitation times and are only explained if the shallower traps have a greater retrapping effect. At very low excitation levels retrapping by largely empty, deeper trapping levels causes T_m to rise.

It may be concluded from the above discussion that the locations of trapping states in the II–VI phosphors appear to be distributed over a range of energies. However, in order to fix the position of a particular type of trap it may be desirable to associate an energy depth with it and under these circumstances the assumption as to the value of s and the method of interpretation should always be indicated with the energy value. If the requirement for the energy depth is not imperative, then the maximum temperature and the heating rate are probably sufficient to identify the trapping state.

4.3.3. *Experimentally Observed Glow Curves*

Zinc sulphide is the II–VI compound whose thermoluminescent properties have been studied in great detail and the results from ZnS

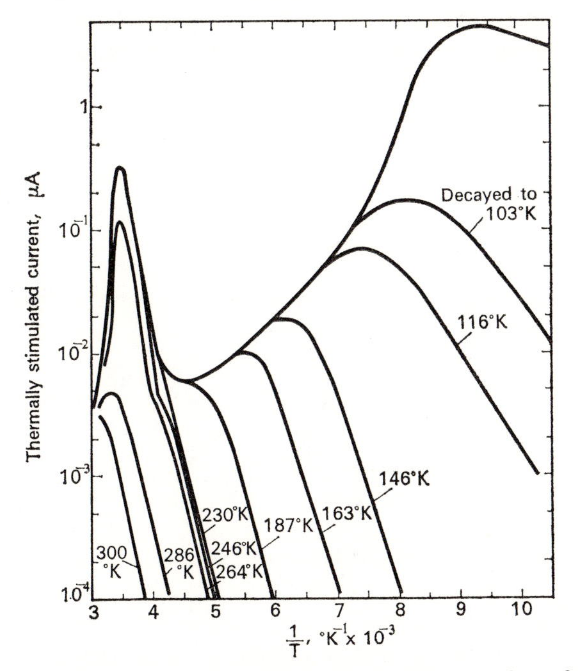

FIG. 4.11. Decayed conductivity glow curves for CdS plotted as a function of reciprocal temperature.[68]

will be used here to exemplify the behaviour of II–VI compounds. Self-coactivation in zinc sulphide has not been the subject of a separate study as was self activation, however, there is evidence of self-coactivating centres from the behaviour of deliberately added coactivators.

It will be assumed unless otherwise stated in the ensuing discussion that the heating rate used for the glow curves is 2°K/sec.

(a) *Halogen coactivation.* Chlorine is an impurity which is present in ZnS as a result of the method of preparation and special measures are necessary to remove it, hence glow curves of zinc sulphide phosphors are frequently characterized by a chlorine originated peak. A single glow peak is observed with both chlorine and bromine coactivation with the latter being the slightly deeper level.[56] In hexagonal ZnS for silver and self-activation the halides provide a peak at $T_m = 180°K$ and for copper activation a peak occurs at $T_m = 185–190°K$. A change to the zinc blende structure results in peaks at approximately

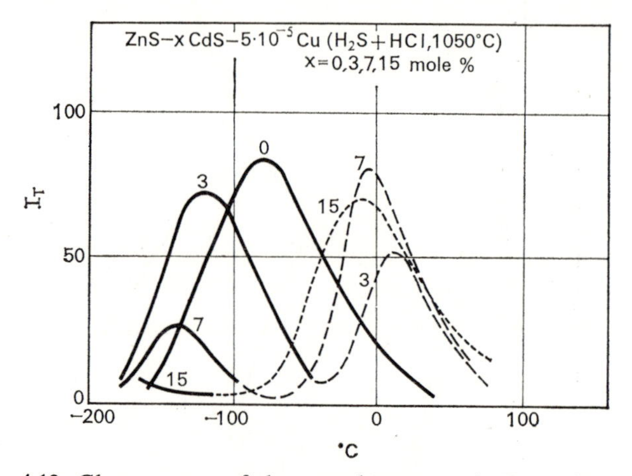

FIG. 4.12. Glow curves of hexagonal structured ZnS–CdS:Cu, Cl phosphors.[56]

20°K lower in temperature. The analysis of both decayed and built-up glow curves indicate that there is a distribution of traps which is particularly pronounced in silver activated ZnS where at $C_o/C_{oo} = 0{\cdot}2$ (section 4.3.2) decayed and built-up T_m values are 211 and 158°K respectively. Heavily copper activated and lightly chlorine coactivated ZnS exhibits a change in the glow emission colour from blue to green as the temperature increases. It is tempting to suggest that the change may indicate two different transfer processes, 4.2.2 (a), although it seems more probable that the change is simply the result of a variation in the emission sensitivity with temperature. The centre thought to be associated with the 180°K peak is simply a halide on a substitutional sulphur site (Cl_S or Br_S).

In copper activated ZnS–CdS alloys the 180°K glow peak of ZnS is observed to shift to lower temperatures with increasing CdS content and occurs below 90°K at 15 mol.% CdS. A new peak is observed at 285°K in the high mol.% CdS alloys and also is found with gold and silver activation. The temperature shifts with composition of the 180 and the 285°K peaks are 10 and 2°K/mol.% respectively. Figure 4.12 illustrates the glow curves of ZnS–CdS:Cu, Cl phosphors. The 285°K peak is generally related to both cadmium and chlorine although no definite model for the centre has been proposed. Cubic ZnS–ZnSe, copper activated alloys exhibit only the single glow peak observed in cubic ZnS at 160°K and the peak has a composition shift of 2·7°K/mol.% ZnSe.

(*b*) *Trivalent metal coactivation.* The behaviour of Al, Sc, Ga and In as coactivators has been summarized by Hoogenstraaten.[56] At low coactivator concentrations relative to the copper activator concentration a poor fluorescence efficiency is found and the emission is temperature quenched at considerably lower temperatures than with halide coactivators. In the poorly fluorescent phosphors a peak occurs at 150°K and this has been associated with self-coactivation resulting from a sulphur vacancy. There is strong evidence for this identification since in conductivity glow studies of single crystal CdS Nicholas and Woods[69] have observed a peak at 150°K, which might be expected to occur at the same temperature as in ZnS if it results from a sulphur vacancy. Greater concentrations of trivalent coactivators cause the self-activated peak to be dwarfed by characteristic peaks for Al, Sc, Ga and In at T_m values of 183, 255, 308 and 368°K respectively. Figure 4.13 compares the glow curves for the four trivalent metal coactivators in copper activated ZnS. With silver activated ZnS both Al and Sc produce similar glow peaks to those in Fig. 4.13 but the peaks for the deep traps resulting from Ga and In do not occur because silver emission is temperature quenched just above room temperature. The single glow peaks observed for appreciable coactivator concentrations has been attributed to the trivalent metal on a substitutional zinc site (Al_{Zn}, Ga_{Zn}).

The form of the glow curves in hexagonal copper activated ZnS–CdS alloys with aluminium coactivation is very similar to those for chlorine coactivation; a composition dependence for the Al_{Zn} glow peak of 6°K/mol.% CdS is obtained. A very broad band peak at approximately 300°K is introduced and this peak shows signs of resolution into two maxima at 7 mol.% CdS. The general indication is that a very broad

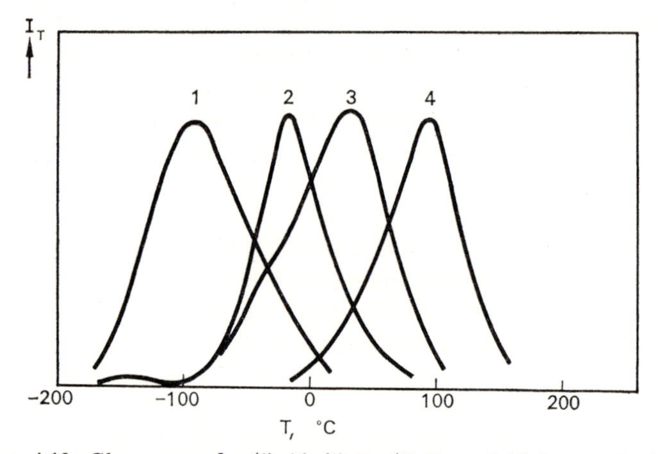

FIG. 4.13. Glow curves for (1) Al, (2) Sc, (3) Ga and (4) In coactivation of copper activated ZnS.[56]

distribution of trap depths exists for this higher temperature peak which is identified with aluminium–cadmium interaction in some way. Scandium and gallium coactivations in the copper activated ZnS–CdS alloys do create additional deeper traps but the 255 and 308°K traps shift with composition at a rate of 8·7 and 8·2°K/mol.% CdS respectively.

Thermal glow measurements commencing at liquid hydrogen temperatures indicate in both Cu–Al and Ag–Al sensitized zinc sulphide phosphors a sharp glow peak at 60°K.[56] The origin of this peak is uncertain but it seems likely to be associated with a natural defect such as a zinc interstitial or sulphur vacancy.

(c) *Oxygen modified phosphors.* Copper activated ZnS when prepared with 2% ZnO and $CaCl_2$ as a flux exhibits three peaks in the glow curve with maximum temperatures at 148, 236 and 320°K. These peaks seem to be typical of oxide doped ZnS phosphors generally and the two higher temperature peaks appear to be created by the introduction of oxygen. The 148°K peak as mentioned above is observed in phosphors with very little added coactivator concentration and it probably results from a sulphur vacancy V_s created to compensate for the added activator. The presence of oxygen on the sulphur sublattice will also tend to create sulphur vacancies because of the size of the oxygen atom and as a result associated sulphur vacancies $[V_s]^2$ have been suggested to account for the 236°K peak. The 320°K peak is thought to be created by the perturbation of the lattice due to the presence of oxygen atoms. Gold activation of ZnS shows similar behaviour to copper activation, while for silver and self-activation the 320°K peak does not occur probably because the quenching is in effect at this temperature. With ZnS–CdS phosphors oxygen creates a single broad emission band for as little as 5 mol. % CdS; under these conditions there would appear to be a complex of trap levels which are difficult to analyse.

(d) *Transition metal behaviour and miscellaneous coactivation.* The emission associated with transition metal elements has been discussed briefly in section 4.2.3 and it has been noted that quenching of copper and silver characteristic emission occurs when transition elements are added.[(47)] Thermoluminescence studies have supplemented the fluorescence studies and suggest the transition metals create deep trapping centres.

Cobalt in ZnS: Cu, Cl creates a glow peak at 385°K which dominates the 180°K peak at a cobalt concentration of 10^{-5} to 1 part ZnS. The cobalt peak is very sharp and suitable for analysis by Garlick and Gibson's method[(60)] which gives $E = 0{\cdot}50 \pm 0{\cdot}04$ eV, $s = 10^6$ sec^{-1}; there is no temperature shift of the peak position with cobalt concentration. For ZnS–CdS:Cu, Cl, Co alloys there is a regular shift of the cobalt peak to lower temperatures with composition at a rate of 8–9°K/mol. % CdS. Slight broadening of the peak occurs probably due to the creation of the Cd–Cl peak mixing with the cobalt peak. Although the cobalt glow is quenched at room temperature in silver activated ZnS, the cobalt peak is observed in the ZnS–CdS alloys at maximum temperatures of 242, 215 and 151°K for 10, 20 and 30 mol. % CdS respectively.

Both nickel and iron tend to reduce the luminescence efficiency in ZnS:Cu, Cl phosphors. Nickel, in particular, causes T_m for the chlorine peak to move to lower temperatures with increasing nickel concentration as the quenching temperature of the phosphor is reduced.

There have been some interesting observations of narrow band thermoluminescent emission in melt grown ZnTe which had been subjected to an anneal in liquid zinc at 850°C for about 60 hr.[67] An extremely narrow glow peak in the red was obtained with a peak at 29°K of half width 2°K and the peak temperature was invariant with respect to change in the heating rate. The peak has been tentatively associated with a bound exciton annihilation of binding energy 0·061 eV. The characteristic glow peak of the anion vacancy occurs at 53°K for a heating rate of 0·14°K/sec.

(e) *Summary.* The interpretation of the thermoluminescence curves in terms of the source of given glow peaks is far from complete. The introduction of appreciable concentrations of elements such as halogens, trivalent elements and cobalt does create single peaks which can be associated with these impurities although only the halogen and cobalt peaks may be identified as simple unassociated substitutional centres. The evidence for a sulphur vacancy as the origin of the 150°K peaks also seems fairly conclusive. The luminescent emission of ZnS phosphors, however, decreases as the purity of the sample increases and hence sets a limitation to the study of low concentrations of impurities which are the ideal conditions for identification of defect states. There is thus an inherent difficulty to identify natural defects from thermoluminescent studies. It is from this point of view that conductivity glow and decay are the more effective techniques for studying trapping states although they do require the phosphor to have an appreciable conductivity. It is perhaps worth adding that the results of conductivity glow measurements on CdS single crystals suggest extremely complex distributions of trapping states without the addition of foreign impurity elements.[68, 69]

Figure 4.14 shows the behaviour of different trapping levels in ZnS–CdS:Cu phosphors. Table 4.2 summarizes the observed glow peak temperatures in ZnS for various coactivator impurities.

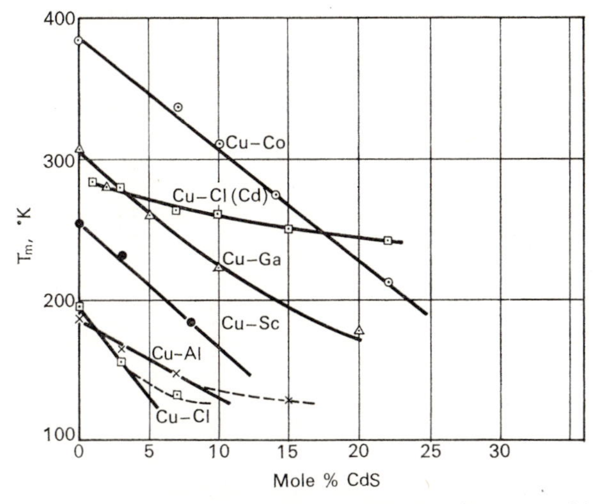

FIG. 4.14. Variation of glow peak temperatures with composition in ZnS–CdS:Cu phosphors.[56]

4.4. Electroluminescence

This branch of luminescent phenomena has until recently been dominated by the study of A.C. electroluminescence in ZnS and other powders. The reasons for such a situation are perhaps fairly obvious in that powders are more easily prepared than single crystals and that the commercial motivation was in large area illumination for which a powder in dielectric suspension was all that was required. However, the upsurge in interest over the past few years in high intensity small area light sources has stimulated greatly the investigation of single-crystal properties and has to some extent redressed the imbalance in knowledge that existed. It is proposed to subdivide the section on electroluminescence on more or less historical grounds into two parts, the first on powders and the second on single crystals and thin films.

TABLE 4.2. SUMMARY OF GLOW PEAKS IN ZnS–Cu PHOSPHORS MEASURED WITH A HEATING RATE OF 2°K/SEC[56]

Coactivator addition	Trapping centre	Peak position T_m (°K)	Trap depth E (eV)	dT_m/dx in $Zn_{1-x}Cd_xS$
None	V_S^+	148	0·27*	Small
Cl	Cl_S^+	185	0·25	10
Br	Br_S^+	190	0·26	—
Al	Al_{Zn}^+	183	0·25	6
Sc	Sc_{Zn}^+	255	0·34	8·7
Ga	Ga_{Zn}^+	308	0·42	8·2
In	In_{Zn}^+	368	0·50	—
	V_S^+	148	0·27*	
O	$(V_S^+)_2$	236	0·44*	Small
	O_s	320	0·59*	
Co	Co_{Zn}	383	0·52	7·8–8·9
Ni	Ni_{Zn}	—	0·58	—
Cd + Cl or Al	Complex	285	0·53*	2
Cl or Al		60	0·1	—

E values calculated from $E = T_m/740$, except where marked with an asterisk when calculated from $E = T_m/540$.

4.4.1. *A.C. Electroluminescence in ZnS-type Powders*

(a) *General experimental observations.* The observation of A.C. electroluminescence is made on a standard three layer sandwich cell illustrated in Fig. 4.15. The cell consists of an upper layer of conducting glass perhaps with a transparent sheet of insulator covering it, a central layer of the phosphor suspended in a dielectric liquid and a copper electrode of a centimeter diameter as the bottom layer. The proportion of phosphor to dielectric liquid that is generally necessary to give satisfactory performance is 1:2 by volume. The relative permittivity of the liquid should not be too great because dielectric loss becomes important although in general a large value of permittivity increases the effective electric field acting on the phosphor particle.

Zinc sulphide is the most studied II–VI compound and has the requirement of high copper concentrations for efficient A.C. electroluminescence. The copper concentrations necessary are in excess of the solubility of copper in zinc sulphide and it is assumed that the copper is precipitated as highly conducting copper sulphide or possibly just as a thin conducting region of copper. It is thought that stacking faults are created in the zinc sulphide particles on cooling from the preparation temperature and the excess copper is precipitated in these faults as copper sulphide or copper. The stacking faults result because ZnS can take two crystal structures and they represent transitions between the two structure types. Lehmann[70] has emphasized this point by deliberately introducing stacking faults into ZnS–CdS alloys with low temperature sulphur bakes and the resultant phosphors were electroluminescent.

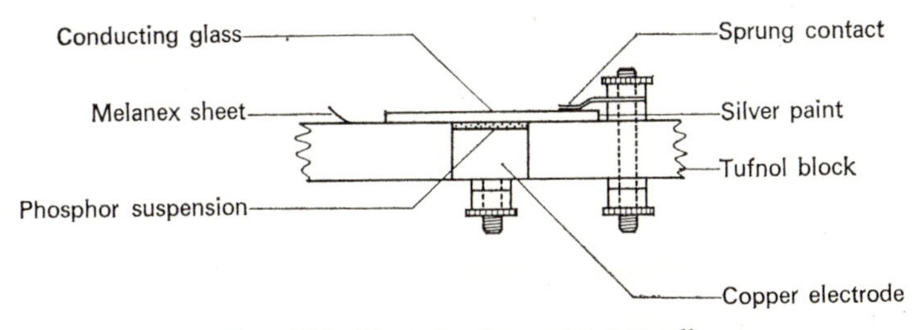

FIG. 4.15. Electroluminescence test cell.

The observation of the electroluminescent emission from particles of ZnS has proved difficult in the past because of the refractive index ($n = 2{\cdot}4$) although techniques have now been developed for such analysis mainly through the efforts of Gillson and Darnell.[71] They produced powder particles with planar faces of millimetre dimensions by firing at high temperatures for long periods. The electroluminescence could be observed from a particle if all but one face was covered with a dark resin. In strong electric fields electroluminescent lines were observed along certain crystal directions. The lines consisted of two comet shaped collinear parts referred to as double comet lines; these are illustrated in Fig. 4.16 for a range of applied electric fields. At low fields the double comet lines reduce to two spots which form the intense part of comet lines at higher fields.

Fischer[72] subsequently showed that the behaviour in the large particle was also typical of that observed from small particles in suspension. He studied the emission of 10 μm particles which were immersed in an optically similar medium of As–S–Br glass to ZnS. The comet line structure was apparent in the whole range of electroluminescent powders investigated. Generally several comet lines were seen even in the smallest particles and comet lines often cut each other at angles of 60°. Detailed observations of the time resolved emission from a comet line for excitation by sinusoidal and square wave pulses indicate that emission occurs when the adjacent electrode swings

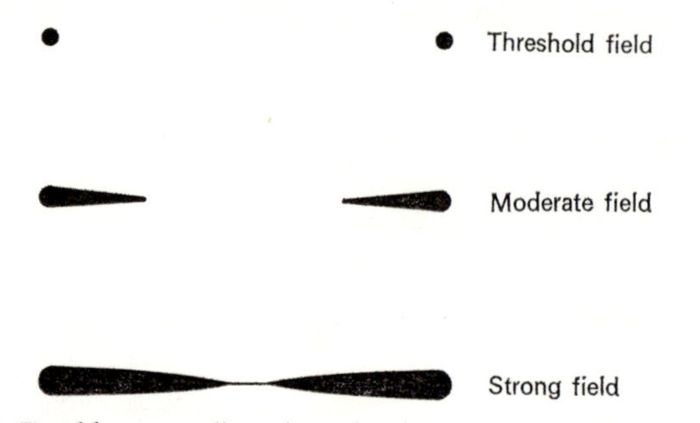

FIG. 4.16. Double comet line electroluminescence in a single particle at weak, medium and strong electric fields.[71]

positive, i.e. the double comet lines do not appear simultaneously but alternately every half cycle. The site of electroluminescence appears to be linear crystal faults, probably dislocation lines decorated with copper sulphide. The electroluminescent emission intensity I from a single comet line for an applied voltage V follows a relationship of the form $I = A \exp(-c/V)$ where A and c are constants. The integrated emission intensity from many comet lines obeys an equation of form $I = A \exp(-c/V^{\frac{1}{2}})$ which it can be shown may be derived from the superposition of many equations of the form $I = A \exp(-c/V)$. The variation of the time integrated emission intensity from single comet lines with frequency, while the applied voltage remains constant, is of a sublinear kind. Frequency saturation of such emission requires a greater frequency at high voltage than at low voltage. Integrated emission from many comet lines has a linear frequency dependence until saturation sets in.

A spectral shift of the electroluminescent emission from ZnS:Cu, Cl phosphors with increasing frequency of excitation from green to blue occurs. Similar behaviour is also observed in the individual comet lines and an explanation has been suggested on the basis of a non-homogeneous copper distribution since at high frequencies the comet lines become narrow and sharp; high copper concentrations favour blue luminescence and close to the stacking faults heavy copper doping is very likely. The emission intensity falls off with decrease in temperature as electrons are more easily trapped, although at considerably lower temperatures an increase in emission occurs because electric field release of electrons from very shallow traps becomes possible. At high temperatures the emission intensity also falls off because of the temperature quenching of the luminescent centres, although strong electroluminescence has been observed up to 450°K for higher frequencies (2–50 kc/s) at which temperature quenching does not have time to dominate the different rate processes.[73]

(b) *Theoretical models*. Many explanations for the electroluminescent behaviour of ZnS-type powder phosphors have been proposed in the form of simple models and these are well reviewed by Fischer[74] and Morehead.[75] However, the situation that exists in a powder suspension is complex and there exists at the present time two fairly detailed models which qualitatively explain most of the observed facts. It is proposed in this text to only discuss these two models.

(i) *Impact ionization model*. Maeda[76] proposed a model in which it was assumed that impact ionization of centres or the lattice by high-energy electrons derived from the conducting needle region occurs in the insulating region of the particle. The light emission occurs when the field is reversed and the electrons return to the ionized centres. A difficulty with the model was that the emission intensities were 180° out of phase with those observed. However, this objection is overcome, if it is assumed that the electrons are field released from shallow traps in the insulating area of the particle and then accelerated by the intense field which results from the conducting region.[74] The impact ionization occurs close to the conductor and the electrons flow into it or towards the positive electrode. On field reversal electrons are released from the conducting needle and recombine with the ionized centres to give radiative emission.

(ii) *Carrier injection model.* Fischer's carrier injection model[77] assumes that the luminescent ZnS particle is a nearly insulating compensated semiconductor with luminescent centres which are created by copper impurity. The dislocation lines take the form of a conducting *p*-type copper sulphide region. On application of an electric field, field intensification occurs at the tips of the conducting needle and electrons and holes are injected simultaneously from opposite ends of the needle into the insulating ZnS. The fact that both electrons and holes are injected means there is no depletion of carriers in the conducting region and hence the process is completely regenerative. The electrons are relatively mobile and only fall into shallow traps whereas the holes are almost immediately trapped in the vicinity of the conducting needle. When the field is reversed the conducting needle again emits electrons at the electrode becoming positive and these electrons recombine radiatively with the trapped holes. Electrons stored at the other end of the microcrystallite flow back through the volume of the crystal and recombine radiatively with trapped holes to give a secondary emission peak as is observed in practice. Figure 4.17 illustrates the energy diagram for the carrier injection model.

These two models fit generally the observed behaviour of the electroluminescent powder, except perhaps that recombination to a limited degree might be expected at both ends of the conducting needle. Even more refined techniques will be necessary to elucidate fine detail in the emission characteristics and the distinct nature of the dislocation sites.

(c) *Particular characteristics of the electroluminescent II–VI powder phosphors.* Strong electroluminescence is observed in ZnS powders doped with concentrations of copper greater than 1 part in 10^3 in molecular proportions. The copper is soluble in the ZnS at the firing temperature (1100°C) and precipitates out as copper sulphide on cooling; the cooling rate has obviously a critical effect on the emission properties of the phosphor and must be particularly slow in the temperature region 600–800°C. The ZnS phosphor particle sizes which result are generally in the range 1 to 50 μm and the particle surfaces are coated with copper or copper sulphide which is usually removed by a wash in cyanide solution. The same trend in the variation of the emission colour is observed with changing coactivator concentration as in the photoluminescent spectra, see Fig. 4.3. If silver is used instead of copper as activator only very weak electroluminescence is obtained; it is assumed that the poor conductivity of precipitated silver sulphide does not

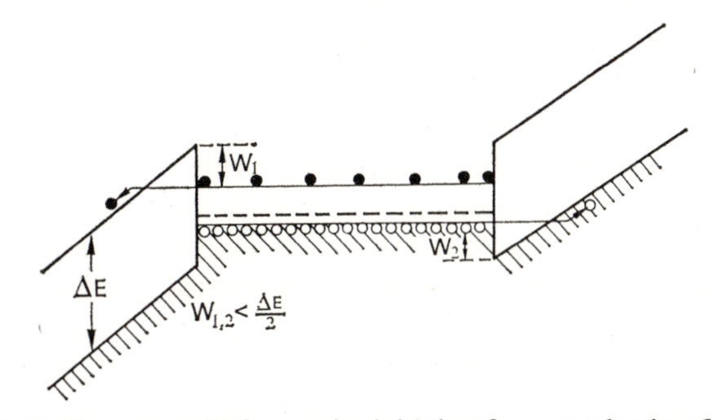

FIG. 4.17. Energy model for carrier injection from conducting Cu_2S into the surrounding ZnS.[77]

create sufficient field intensification to give strong electroluminescence in the powders. The introduction of manganese or rare earths in addition to copper as activators in ZnS results in electroluminescent emission which is characteristic of manganese or the rare earth. However, if copper is not present only very weak emission is observed.

ZnSe phosphors doped with copper and aluminium or chlorine exhibit broadly similar properties to ZnS. The brightness level in ZnSe is considerably less at room temperature than in ZnS because the presence of selenium on the anion site causes a reduction in the energy separating the valence band and the activator level and thus enhances thermal quenching effects. As with ZnS, thermal quenching effects can be minimized by studying the ZnSe phosphor at high frequencies. Gelling and Haanstra[78] have investigated the spectral nature of the electroluminescent emission from ZnSe at room temperature and have observed a peak at 1·94 eV (645 mμm) with a half width of 0·30 eV; there appears to be negligible shift of the peak energy with increasing frequency. Substitution of Cd for Zn on the anion sub-lattice gave more serious temperature quenching effects although at high frequencies the quantum efficiency was similar to that of ZnSe phosphors. The emission peak energy at room temperature had shifted to 1·8 eV for alloys containing 8 mol.% CdSe. CdS phosphors do not exhibit A.C. electroluminescence and this fact is usually a result of the absence of stacking faults in the wurtzite structure of CdS. Electroluminescence in the ZnS–CdS system only occurs in alloys with CdS concentrations up to 20% above which the zinc blende structure ceases to be stable at the preparation temperature;[71, 79, 80] this represents further support for the models that require stacking faults to be present for electroluminescence in powders to occur. ZnTe with a smaller energy gap (2·26 eV at 300°K) than either ZnSe or CdS does not exhibit visible electroluminescence at room temperature because of strong thermal quenching effects; however, if ZnTe is cooled to liquid nitrogen temperatures weak emission in the yellow spectral region occurs.[75] The remainder of the II–VI compounds in powder form have not been observed to emit in the visible spectral region to which the layer sandwich cell investigations are usually restricted. Generally alloys of the electroluminescent compounds which have stacking faults are also electroluminescent; the ZnS–ZnSe system is perhaps the most useful set of alloys in that the electroluminescent emission colour varies from blue to red across the composition range.[81, 82] For further details of the visible emitting alloys formed between the II–VI compounds the reader is referred to the article by Wachtel.[79]

4.4.2. *Electroluminescence in Single Crystals and Thin Films*

The attraction of injection electroluminescence in single crystals and thin films has stimulated much effort in the preparation of II–VI compounds in suitably conducting forms. II–VI compounds have most of the desirable characteristics for efficient visible electroluminescence such as direct energy gap, efficient energy conversion in photoluminescence, emission energies over the whole visible spectrum if alloys are used, and chemical stability. However, all the compounds with energy gaps in excess of 2 eV have one unfortunate feature that they are subject to strong compensation effects because of the partial ionic character in the bond type and hence have poor conductivities; conductivities can be improved by the introduction of impurities or the removal of the vacancies by annealing in the appropriate elemental vapour, although generally conductivity of one carrier type only can be obtained. Thus very few *p–n* homojunctions have been produced in the larger energy gap II–VI compounds, although some success has been achieved with alloys and the smaller energy gap (1·6 eV) compound, CdTe. Heterojunctions and other approaches have been required to observe electroluminescence generally in the larger energy gap II–VI compounds. This section will look at the different approaches to the problems of electroluminescence in microcrystalline II–VI compounds.

(a) *P–n homojunctions.* Cadmium telluride is the only II–VI compound with an appreciable energy gap ($> 1{\cdot}0$ eV) which can be prepared with both *p*- and *n*-type conductivities. CdTe in many respects bears favourable comparison with GaAs in which efficient electroluminescence at a *p–n* junction is a well established and applied phenomenon. Heavily doped CdTe can have resistivities of 10^{-4} and 5×10^{-3} Ω-cm for *n*- and *p*-type material respectively;[83] the dopants used are aluminium (*n*) and phosphorus (*p*). Mandel and Morehead[84] have produced *p–n* junctions in CdTe which emit under forward bias conditions at 77°K with the electroluminescent emission peaking at 850 mμm and having an external quantum efficiency of 12%; the internal quantum efficiency is thought to approach 100%. The electroluminescent emission is largely quenched at 300°K as the shallow acceptor level, at which recombination is thought to occur, lies only 0·06 eV above the valence band. Electroluminescent emission of a coherent nature from CdTe has not yet been observed and the reasons for this are thought to be the high contact resistance to CdTe and lack of planarity in the junction region.

The obvious extension of the limited success achieved with CdTe was to look at alloys formed with CdTe. ZnTe matches CdTe well from atomic size considerations and it can be obtained with a *p*-type resistivity as low as $10^{-3}\Omega$-cm; also it does not suffer from the contact problems of CdTe.[85] *P–n* junctions in $Zn_{1-x}Cd_xTe$ alloys are produced by simultaneous diffusion of zinc and phosphorus into a CdTe:Al crystal in a cadmium atmosphere at 850°C. The zinc which diffuses much faster than the phosphorus converts the bulk of the CdTe crystal into an *n*-type $Zn_{1-x}Cd_xTe$ alloy with a *p*-type skin doped with phosphorus. Emission in the red occurs at 77°K for x values between 0·58 and 0·70 and external quantum efficiencies of 4% have been obtained. The application of high current pulses ($8 \times 10^8 A/m^2$ at 4°K) has failed to produce stimulated emission which has again been ascribed to non-planarity of the junction layer. With high ZnTe concentrations the quantum efficiency falls off rapidly as the *n*-type resistivity increases.

Yamamoto *et al.*[86] have successfully tried an alternative course to exploit the amphoteric nature of CdTe and at the same time achieve emission in the visible. Solid solutions of $Cd_{1-x}M_xgTe$ were prepared by direct synthesis and single crystals were grown from the polycrystalline alloy by a Stockbarger–Bridgmann technique (section 2.4.2). MgTe, although it is itself unstable in air, when in solid solution with CdTe is stable in moist air to $x \sim 0·75$. *P–n* junctions have been made by simultaneous diffusion of P and Mg into *n*-type CdTe crystals with junction areas of 1 mm^2. Visible electroluminescent emission at room temperature has been observed in alloys for which $x = 0·22$ and 0·35 with peak energies of 1·65 and 1·85 eV respectively. The location of the emission is thought to be in the space charge region and the threshold current density for visually observable emission in 5 mA/mm^2.

The ZnSe–ZnTe system is one which has produced some interesting results in respect of visible emitting *p–n* junctions. ZnSe and ZnTe only exhibit appreciable conductivities when they are *n*- and *p*-type respectively; however, there is likely to exist a composition range in their alloys over which both conductivity types may be obtained. Two approaches have been made to the preparation and study of ZnSe–ZnTe homojunctions by Fischer[83] and Aven.[87]

(1) Fischer[83] used an *n*-type ZnSe crystal and grew epitaxially a

p-type ZnTe layer onto it. Under these growth conditions a composition gradient was thought to exist at the p–n junction and two separate n-type ZnSe and p-type ZnTe regions existed on either side of the junction. Strictly speaking this junction is not a homojunction since the base lattices on opposite sides of it are different, though in practice its behaviour is like that of a homojunction and it is often called a quasihomojunction. The emission was observed to come from the n-type ZnSe region close to the junction and suggested that possibly hole rather than electron injection was favoured.

(2) Aven[(87)], on the other hand, started with a p-type $ZnSe_x Te_{1-x}$ crystal grown by vapour transport and diffused in n-type impurity by immersing the crystal in liquid $Zn_{0.99} Al_{0.01}$ alloy at 900°C. The compositions at which useful n- and p-type conductivities were obtained ranged from $x = 0{\cdot}3$ to $0{\cdot}6$. The electroluminescent emission at 80°K observed from a p–n junction for $x = 0{\cdot}36$ was dominated by a peak at 1·98 eV (625 mμm) with a half width of 0·12 eV. The threshold for emission at 78°K was 2·7 V at a current of 2×10^{-8} A for an active junction area of 4 mm^2. External quantum efficiencies of 13% at 78°K and 15% at 80°K occurred for current levels of 10^{-4} A. As the temperature increases, holes are removed from the shallow acceptor centres located 0·06 eV above the valence band and are retrapped in deeper centres at which radiative recombination does not occur. Aven concludes from these and other current dependent observations that radiative recombination occurs in the space-charge region of the p–n junction and that the internal quantum efficiency is close to 100%. More recent investigations of the light emission from p–n junctions in high resistivity $ZnSe_xTe_{1-x}$ crystals suggest that the mechanism for charge transport and emission is self-induced photoconductivity on both the p- and n-type sides of the junction.[(88)]

(*b*) *P–n heterojunctions.* Aven and Cusano,[(89)] in an analysis of heterojunctions formed between copper chalcogenides and II–VI compounds, have observed injection electroluminescence in Cu_2Se–ZnSe and Cu_2S–ZnS heterojunctions. It is interesting that in some earlier work these same authors had observed the reverse effect of photovoltaic generation in Cu_2Te–CdTe and Cu_2S–CdS heterojunctions, section 7.7. Information on the Cu_2Se–ZnSe heterojunction is the best documented and the experimental evidence suggests that the mismatch between the energy gap of Cu_2Se and ZnSe has been overcome by the presence of an intermediate, semiconducting layer of ZnSe. The intermediate layer is created by the diffusion of copper through the cleaved (110) junction surface of the ZnSe substrate and subsequent compensation of donor impurities. Electrically the semi-insulating region acts as an impedance to the reverse flow of electrons from ZnSe into Cu_2Se until sufficient voltage has been built up at the valence band edge for hole injection. In the absence of an intermediate layer no electroluminescence is observed, although at the other extreme

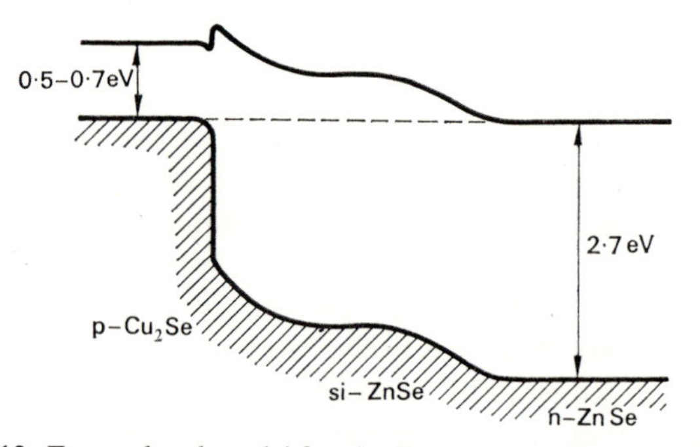

FIG. 4.18. Energy band model for the ZnSe–Cu_2Se heterojunction.[89]

the insulating layer must not be too thick otherwise the electron flow is insufficient for significant emission to occur. An energy band model, Fig. 4.18, has been proposed which explains satisfactorily the current–voltage characteristic of the emitting junction with two different exponential rise regions from thresholds of 0·5 and 1·2 eV.[89] The threshold voltages at 300 and 77°K for electroluminescent emission in the Cu_2Se–ZnSe heterojunctions are 1·4 and 1·5 V respectively. The spectral division of the electroluminescence at 77°K is into three peaks at 1·96 eV (red), 2·35 eV (green) and 2·68 eV (blue–green). These peak emission values are broadly in agreement with those obtained from the photoluminescence studies discussed in section 4·2; the 1·96 eV and

2·35 eV peaks of ZnSe are the equivalent of the copper green and blue emission in ZnS. Electroluminescence emission has also been observed from Cu_2S–ZnS heterojunctions, although analysis of the results has not been possible because the effects of the poor ohmic contacts made with the ZnS cannot be assessed.

There has been a suggestion by Fischer[(74)] that the non-planarity of a junction between the copper chalcogenide and ZnS or ZnSe may lead to carrier injection by other means. The two obvious injection mechanisms are those considered for powders [section 4.4.1 (b)] in which narrow conducting regions of Cu_2Se project into ZnSe and cause field intensification. The observation of electroluminescence under D.C. conditions is possible, if carrier depletion at recombination creates an internal reverse field and thus perpetuates the luminescence process.

(c) *Metal–semiconductor contacts.* Metal contacts evaporated onto single crystals or crystalline thin films provided obvious configurations for the investigation of conducting and electroluminescent properties. The nature of the contacts, however, are often very uncertain and show changes with time or operation of the configuration. As a result metal–semiconductor contacts in various forms have provided a multiplicity of explanations for the observed electroluminescence.

The investigations of Thornton[(90–92)] on ZnS:Cu, Mn, Cl thin films evaporated onto a conducting glass substrate and with an evaporated upper electrode of aluminium were among the first to come down in favour of a mechanism of carrier injection other than avalanche injection resulting from impact ionization. Thornton observed both D.C. and A.C. electroluminescence in these thin ZnS films at 2·0 V, which does not give high enough electron energies to impact ionize impurity centres or the lattice. It was assumed that minority carrier injection occurred at the aluminium–zinc sulphide interface rather in the same way as at a heterojunction. Subsequent investigations[(93)] have suggested that a forming time is required before electroluminescence is observed and then emission appears only at the aluminium electrode when it is negative. The deductions made from these observations are that an insulating barrier of Al_2O_3

is necessary at the Al–ZnS interface before emission and tunnel injection of holes occur with resultant carrier recombination at impurity centres. The operating lifetime of these films is only of the order of tens of hours which is not unexpected if the growth of Al_2O_3 continues and eventually prevents the carrier flow. Figure 4.19 illustrates a possible energy band model for hole injection through an insulating film. Figure 4.19 (a) suggests that in the region of the barrier the ZnS becomes more

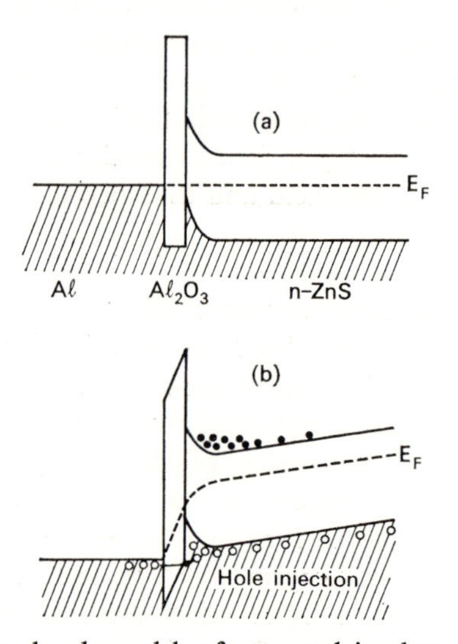

FIG. 4.19. Energy band model of a metal–insulator–semiconductor contact, (a) under thermal equilibrium conditions, (b) with an applied voltage sufficient to stimulate hole injection.

insulating, possibly through the influence of surface states which compensate more strongly the donor states. The applied field condition in Fig. 4.19 (b) is such that holes are tunnel injected for energies less than that of the band gap; the holes accumulate at the insulator–ZnS interface as do electrons and recombination thus gives rise to emission in the region of the anode. With this model electron extraction at the anode is negligible except for very high fields. Similar results to the above were obtained for ZnS thin films with an insulating layer of silicon monoxide to separate the phosphor from an evaporated copper electrode; in the absence of the silicon monoxide layer no electroluminescence was observed at all. The level of efficiency achieved with this form of electroluminescence is not high and probably does not exceed 0·1%. Antcliffe[94, 95], in studies on manganese activated ZnS and ZnSe thin films with similar electrodes to those used by Goldberg and Nickerson[93], did not observe electroluminescence for voltages less than 12 V; high resistances were observed for these films ($10^8 \Omega$) and it is conceivable that the voltage drop does not occur wholly at the metal–semiconductor interface but largely over the ZnS or ZnSe film, hence the behaviour should be similar to that of a powder cell.

Investigation of cleaved single crystals of CdS with a 10^{-8} m silicon monoxide layer to separate the crystal from a thick evaporated layer of gold tend also to suggest tunnel injection of minority carriers.[96] Systematic study of both two and three electrode structures with CdS have indicated a minimum of 1·3 V for electroluminescence to occur at 77°K. Green edge emission was observed at the surface of the crystal closest to the anode and was polarized preferentially perpendicular to the *c*-axis of the crystal. It is possible that the emission observed from supposedly *p*-type, copper-doped ZnSe crystals alloyed to an indium contact with a threshold of 1·1 V may also result from tunnel injection.[97] Minority carrier injection electroluminescence at room temperature has also been reported in diodes formed between indium and *p*-type ZnTe, although no indication of the voltage threshold is given;[98] the emission was orange in colour and always occurred at the anode. Electroluminescence had also been observed from ZnTe diodes at 77°K and quantum efficiencies of 2% have been achieved at very high current levels ($\sim$ 5000 A/cm^2).[99]

It is clear from the above discussions that tunnel injection is a likely mechanism for electroluminescence at metal–semiconductor interfaces with an intermediate insulating layer. In the absence of the intermediate layer other mechanisms become operative and of these avalanche injection by impact ionization to be described below, is favoured. One interesting comment on the variety of mechanisms is illustrated by some observations on ZnS:Cu, Cl films of the order of millimetres thick.[83] At low fields fine spots were observed at different points in the film but as the field increases so the emission spreads uniformly throughout the film. To conclude this section mention should perhaps be made of the fact that very little use has been made of high work function metals as hole injectors into the *n*-type II–VI compounds.

(d) *Avalanche injection.* If sufficiently strong localized field strengths can be built up in a semiconductor, carriers can be accelerated to energies at which they can impact ionize defect centres and the lattice. A reverse biased *p–n* junction often represents a suitable condition for carrier

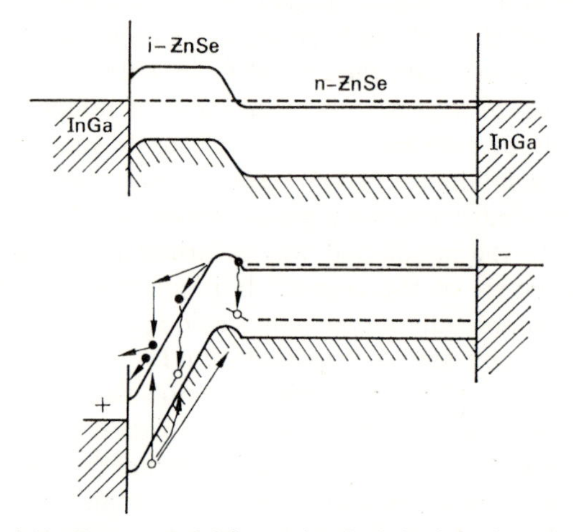

FIG. 4.20. Energy model for avalanche injection of carriers.[83]

acceleration since the applied field enhances the effect of the built-in field at the depletion layer. Although no minority carriers are injected across the depletion layer, those carriers (both types) in existence are accelerated to impact ionize defect centres or the lattice and carrier avalanches are created with a resultant growth in radiative recombination. Similarly an insulating section adjacent to a conducting section of a crystal can create conditions for field acceleration of carriers and subsequent avalanche injection. Figure 4.20 illustrates the energy diagram for the latter example with an electric field applied. A much fuller discussion of the conditions required for avalanche injection leading to electroluminescence is given by Thornton.[100]

The observation of electroluminescence in thin films of ZnS:Mn governed by impact ionization was indicated in section 4.4.2 (c). The chemical requirement of the manganese activated ZnS and ZnSe thin films for the occurrence of electroluminescence was that excess zinc or manganese must be present. This fact suggests that regions of different resistivity level existed in the films and that high resistivity areas could create high local fields where trapped carriers are field released, accelerated and lead to avalanche injection.

Aten and Haanstra[101] have obtained electroluminescence from tellurium doped CdS crystals with a peak in the orange (2·08 eV) at 77°K. The Ag–CdS–In sandwich used in the investigation exhibited emission over large areas and the high voltages required are suggestive of avalanche injection. Tsujimoto and Fukai[102] have studied the current–voltage characteristics and the electroluminescent emission from ZnSe–ZnTe heterojunction diodes under forward and reverse bias conditions at room temperature. Good rectification properties are observed and a threshold of approximately 20 V is required for electroluminescence under reverse-biased conditions. A broader emission spectrum is found at reverse bias than at forward bias, as might be expected since hot carriers are involved in recombination transitions and it is quite likely that intraband transitions help to extend the spectrum to longer wavelengths. Although ZnTe can be made strongly *p*-type by doping with phosphorus or arsenic which form shallow acceptor levels,[75] visible electroluminescence is not observed at room temperature because of temperature quenching effects. Kennedy and Russ[103] have overcome the difficulty of temperature quenching by the introduction of oxygen as a deep acceptor state and then electroluminescence occurs at room temperature with a dominant peak at 696 mμm (external quantum efficiency of 0·1–0·01 %). At 77°K the peak shifts to 685 mμm and the efficiency increases by a factor of ten. The threshold for emission, which occurs over the bulk of the ZnTe, is 4·5 V; the device structure used had a diffused indium contact and soldered In–Ag contact and the brightest emission occurs when the diffused indium contact is negative with respect to the *p*-type ZnTe. The conclusions drawn by these investigators is that the most probable mechanism for electroluminescence is impact ionization with the accompanying avalanche injection and the recombination occurs at an acceptor level located 0·45 eV above the valence band.

References

1. Garlick, G. F. J., *Luminescence of Inorganic Solids*, pp. 685–731 (P. Goldberg ed.). Academic Press, New York, 1965.
2. Warschauer, D. M., and Reynolds, D. C., *J. Phys. Chem. Solids* **13,** 251–6 (1960).
3. Obrikat, D., Meyer, K., and Polly, F., *Phys. Stat. Sol.* **22,** K123–6 (1967).
4. Schön, M., *Z. Physik.* **119,** 463–71 (1942).
5. Klasens, H. A., *J. Electrochem. Soc.* **100,** 72–80 (1953).
6. Lambe, J., and Klick, C. C., *Phys. Rev.* **98,** 909–14 (1955).
7. Prener, J. S., and Williams, F. E., *J. Electrochem. Soc.* **103,** 342–6 (1956).
8. Van Gool, W., *Philips Res. Repts. Suppl.* **3,** 1–119 (1961).
9. Drozd, L., and Levshin, V. L., *Optika Spektrosk.* **10,** 408–13 (1961).
10. Halsted, R. E., Aven, M., and Coghill, H. D., *J. Electrochem. Soc.* **112,** 177–81 (1965).
11. Holton, W. C., DeWit, M., and Estle, T. L., *International Symposium on Luminescence*, pp. 454–9. Karl Theimig KG, Munich, 1965.
12. Prener, J. S., and Williams, F. E., *J. Chem. Phys.* **25,** 361 (1956).
13. Koda, T., and Shionioya, S., *Phys. Rev.* **136A,** 541–55 (1964).
14. Blount, G. H., Phipps, P. B. P., and Bube, R. H., *J. Appl. Phys.* **38,** 4550–2 (1967).
15. Lehmann, W., *J. Electrochem. Soc.* **114,** 83–87 (1967).
16. Gashurov, G., and Banks, E., *J. Electrochem. Soc.* **114,** 1143–8 (1967).
17. Van Gool, W., and Cleiren, A. P., *Philips Res. Repts.* **15,** 238–53 (1960).
18. Curie, D., and Prener, J. S., *Physics and Chemistry of II–VI Compounds*, pp. 433–85 (M. Aven and J. S. Prener eds.). North Holland, Amsterdam, 1967.
19. Lehmann, W., *J. Electrochem. Soc.* **113,** 449–55 (1966).
20. Curie, D., *Compt. Rend.* **258,** 3269–71 (1964).
21. Browne, P. F., *J. Electronics* **2,** 1–16 (1956).
22. Shionoya, S., Era, K., and Washizawa, W., *J. Phys. Soc. Japan* **21,** 1624 (1966).
23. Lempicki, A., *J. Electrochem. Soc.* **107,** 404–9 (1960).
24. Shionoya, S., Kobayashi, Y., and Koda, T., *J. Phys. Soc. Japan* **20,** 2046–53 (1965).
25. Birman, J. L., *J. Electrochem. Soc.* **107,** 409–17 (1960).
26. Morehead, F. F., *J. Phys. Chem. Solids* **24,** 37–44 (1963).
27. Fonger, W. H., *Phys. Rev.* **137A,** 1038–48 (1965).
28. Aven, M., and Potter, R. M., *J. Electrochem. Soc.* **105,** 134–40 (1958).

29. DIELEMAN, J., *et al.*, *Philips Res. Repts.* **19,** 311–18 (1964).
30. MEIJER, G., *J. Phys. Chem. Solids* **7,** 153–8 (1958).
31. BROSER, I., and SCHULTZ, H. J., *J. Electrochem. Soc.* **108,** 545–8 (1961).
32. POTTER, R. M., and AVEN, M., *Bull. Am. Phys. Soc.* **4,** 227 (1959).
33. BROSER, I., and FRANKE, K. H., *J. Phys. Chem. Solids* 1013–20 (1965).
34. BRYANT, F. J., and COX, A. F. J., *Brit. J. Appl. Phys.* **16,** 463–9 (1965).
35. LEHMANN, W., *J. Electrochem. Soc.* **113,** 788–92 (1966).
36. AVINOR, M., *J. Electrochem. Soc.* **107,** 608–11 (1960).
37. BUBE, R. H., *Phys. Rev.* **90,** 70–80 (1953).
38. MCCLURE, D. S., *J. Chem. Phys.* **39,** 2850–5 (1963).
39. LANGER, D., and IBUKI, S., *Phys. Rev.* **138A,** 809–15 (1965).
40. GUMLICH, H. E., MOSER, R., and NEWMANN, E., *Phys. Stat. Sol.* **24,** K13–16 (1967).
41. GUMLICH, H. E., *et al.*, *J. Chem. Phys.* **44,** 3929–34 (1966).
42. AVINOR, M., and MEIJER, G., *J. Phys. Chem. Solids* **12,** 311–15 (1960).
43. MEIJER, G., and AVINOR, M., *Philips Res. Repts.* **15,** 225–37 (1960).
44. BRIL, A., and VAN MEURS-HOCKSTRA, W., *Philips Res. Repts.* **17,** 280–2 (1962).
45. ALLEN, J. W., *Physica* **29,** 764–8 (1963).
46. GARLICK, G. F. J., *J. Phys. Chem. Solids* **8,** 449–57 (1959).
47. SLACK, G. A., and O'MEARA, B. M., *Phys. Rev.* **163,** 335–41 (1967).
48. ATEN, A. C., HAANSTRA, J. H., and DEVRIES, H., *Philips Res. Repts.* **20,** 395–403 (1965).
49. DIELEMAN, J., DE JONG, J. W., and MEYER, T., *J. Chem. Phys.* **45,** 3174–84 (1966).
50. LANDER, J. J., *J. Phys. Chem. Solids* **15,** 324–34 (1960).
51. IBUKI, S., and LANGER, D., *J. Chem. Phys.* **40,** 796–808 (1964).
52. ANDERSON, W. W., *Phys. Rev.* **136A,** 556–66 (1964).
53. KINGSLEY, J. D., PRENER, J. S., and AVEN, M., *Phys. Rev. Letters* **14,** 136–8 (1965).
54. KOMIYA, H., *J. Phys. Soc. Japan* **24,** 216 (1968).
55. GARLICK, G. F. J., *Luminescent Materials*, pp. 27–43, 112–21. Oxford University Press 1949.
56. HOOGENSTRAATEN, W., *Philips Res. Repts.* **13,** 515–693 (1958).
57. CURIE, D., *Luminescence in Crystals*, pp. 142–89, Methuen, 1963.
58. BUBE, R. H., *Phys. Rev.* **99,** 1105–16 (1955).
59. HAERING, R. R., and ADAMS, E. N., *Phys. Rev.* **117,** 451–4 (1960).
60. GARLICK, G. F. J., and GIBSON, A. F., *Proc. Phys. Soc.* **60,** 574–90 (1948).
61. HAAKE, C. M., *J. Opt. Soc. Amer.* **47,** 649–52 (1957).
62. GROSSWEINER, L. I., *J. Appl. Phys.* **24,** 1306–7 (1953).
63. BOOTH, A. H., *Can. J. Chem.* **32,** 214–15 (1954).
64. HILL, J., and SCHWED, P., *J. Chem. Phys.* **23,** 652–8 (1955).
65. HALPERIN, A., and BRANER, A. A., *Phys. Rev.* **117,** 408–22 (1960).
66. BUBE, R. H., *J. Appl. Phys.* **35,** 3067–9 (1964).
67. TANIMIZU, S., and OTOMO, Y., *Phys. Letters* **25A,** 744–5 (1967).
68. BUBE, R. H., *et al.*, *J. Appl. Phys.* **37,** 23–31 (1966).
69. NICHOLAS, K. M., and WOODS, J., *Brit. J. Appl. Phys.* **15,** 783–95 (1964).
70. LEHMANN, W., *J. Electrochem. Soc.* **110,** 759–66 (1963).
71. GILLSON, J. L., and DARNELL, F. J., *Phys. Rev.* **125,** 149–58 (1962).

72. FISCHER, A. G., *J. Electrochem. Soc.* **109,** 1043–9 (1962).
73. MOREHEAD, F., *J. Electrochem. Soc.* **105,** 461 (1958).
74. FISCHER, A. G., *Photoelectronic Materials and Devices*, pp. 1–99 (S. Larach ed.). Van Nostrand, New York, 1958.
75. MOREHEAD, F., *Physics and Chemistry of II–VI Compounds*, pp. 1–99 (M. Aven and J. S. Prener eds.). North Holland, Amsterdam, 1967.
76. MAEDA, K., *J. Phys. Soc. Japan* **15,** 2051–3 (1960).
77. FISCHER, A. G., *J. Electrochem. Soc.* **110,** 733–48 (1963).
78. GELLING, W. G., and HAANSTRA, J. H., *Philips Res. Repts.* **16,** 371–5 (1961).
79. WACHTEL, A., *J. Electrochem. Soc.* **107,** 682–8 (1960).
80. BALLENTYNE, D. W. G., and RAY, B., *Physica* **27,** 337–41 (1961).
81. MOREHEAD, F., *J. Phys. Chem. Solids* **24,** 37–44 (1963).
82. BALLENTYNE, D. W. G., DEV, I., and RAY, B., *Physica* **30,** 223–8 (1964).
83. FISCHER, A. G., *Luminescence in Inorganic Solids*, pp. 559–602 (P. Goldberg ed.). Academic Press, New York, 1966.
84. MANDEL, G., and MOREHEAD, F. F., *Appl. Phys. Letters* **4,** 143–5 (1964).
85. MOREHEAD, F. F., and MANDEL, G., *Appl. Phys. Letters* **5,** 53–54 (1964).
86. YAMAMOTO, R., *et al., Jap. J. Appl. Phys.* **6,** 537 (1967).
87. AVEN, M., *Appl. Phys. Letters* **7,** 146–8 (1965).
88. AVEN, M., and GARWACKI, W., *J. Appl. Phys.* **38,** 2302–12 (1967).
89. AVEN, M., and CUSANO, D. A., *J. Appl. Phys.* **35,** 606–11 (1964).
90. THORNTON, W. A., *Phys. Rev.* **116,** 893–4 (1959).
91. THORNTON, W. A., *Phys. Rev.* **122,** 58–59 (1961).
92. THORNTON, W. A., *J. Appl. Phys.* **33,** 3045–8 (1962).
93. GOLDBERG, P., and NICKERSON, J. W., *J. Appl. Phys.* **34,** 1601–8 (1963).
94. ANTCLIFFE, G. A., *Brit. J. Appl. Phys.* **16,** 1467–75 (1965).
95. ANTCLIFFE, G. A., *Brit. J. Appl. Phys.* **17,** 327–36 (1966).
96. JAKLEVIC, R. C., *et al., Appl. Phys. Letters* **2,** 7–9 (1963).
97. LOZYKOWSKI, H., *Czech. J. Phys.* **B13,** 164–71 (1963).
98. WATANABE, N., USUI, S., and KANAI, Y., *Jap. J. Appl. Phys.* **3,** 427–8 (1964).
99. MIKSIC, M., *et al., Phys. Letters* **11,** 202–3 (1964).
100. THORNTON, P. R., *The Physics of Electroluminescent Devices*, pp. 83–106. Spon, London, 1967.
101. ATEN, A. C., and HAANSTRA, J. H., *Phys. Letters* **11,** 97–98 (1964).
102 TSUJIMOTO, Y., and FUKAI, M., *Jap. J. Appl. Phys.* **6,** 1024–5 (1967).
103. KENNEDY, D. I., and RUSS, M. J., *J. Appl. Phys.* **38,** 4387–90 (1967).

CHAPTER 5

PHOTOCONDUCTIVITY AND ASSOCIATED BEHAVIOUR

IT HAS been decided to treat photoconductivity separately from transport properties because of the considerable emphasis in the study of II–VI compounds that has been placed on photoconductive behaviour. Photoconductivity in solids is a subject which has been studied for close on a hundred years, although any understanding that exists of photoconductive processes has only been achieved in recent times. The development of the theory of photoconductors is seen to be related to the sudden upsurge in commercial activity when photoconductive devices became competitive with existing devices in fields such as radiation detection and photography. The general background to photoconductivity has been reviewed in several excellent articles by Bube[1–3] and Moss;[4, 5] it is proposed in this chapter to summarize the background ideas and then to look at the behaviour of particular II–VI compounds and their alloys. The discussion is restricted principally to photoconductivity in its most obvious form and the reader is referred to other works for the basic ideas on photostatic charge effects and photoelectromagnetic effects.[5–7]

5.1. Background Ideas of Photoconductivity and their Relevance to II–VI Compounds

The absorption of radiation of suitable frequency by any compound leads to the creation of free electron–hole pairs. In the II–VI compounds the holes are generally very much less mobile than the electrons and consequently are captured more readily by defect centres; in their turn electrons will also be captured at defect centres, but it is the length of time these electrons spend in the conduction band that determines the usefulness of the compound as a photosensitive material. The nature of the defect centres is critical in determining the behaviour of the majority carriers in the n-type II–VI compounds. Compounds, in which defects behave simply as trapping centres in thermal equilibrium with the conduction or valence bands, do not exhibit strongly photosensitive properties. In order to achieve high photosensitivity the important defect centre, which is of an acceptor type in an n-type material, must behave as a recombination centre and not be greatly affected by thermal equilibrium considerations. The experimental evidence on halogen-doped CdS indicates that highly photosensitive material has its behaviour controlled by compensated natural acceptor defect centres. The property of the centre is that its capture cross-section for free holes is of the order of 10^5 times greater than its subsequent capture cross-section for free electrons. The relative values of the transit time of the electron across the crystal and the electron lifetime in the conduction band determine the actual photoconductive gain. The upper limit of gain is set by the condition when more electrons are being injected from the ohmic contact than are being created by photoexcitation.

5.1.1. *Photosensitivity*

The conductivity of an n-type semiconductor is given by

$$\sigma = ne\mu_n \tag{5.1}$$

where n is the density of free majority carriers and μ_n is the carrier mobility. If a change in conductivity, $\Delta\sigma$, results on illumination of the material, then it can be described by

$$\Delta\sigma = \Delta ne\mu_n + ne\Delta\mu_n \tag{5.2}$$

Since the number of free carriers is related to the excitation rate f and the free carrier lifetime τ by

$$n = f\tau \tag{5.3}$$

$$\Delta n = \Delta f\tau + f\Delta\tau$$

Therefore

$$\Delta\sigma = e\mu_n\tau\Delta f + e\mu_n f\Delta\tau + ef\tau\Delta\mu_n \tag{5.4}$$

thus the photosensitive change in conductivity can be related to three factors:

(1) $e\mu_n \tau \Delta f$, which simply relates to the change in excitation rate and would seem to be the obvious concept of photoconductivity.

(2) $e\mu_n f \Delta\tau$ is a term that suggests the carrier lifetime might change with excitation conditions. A positive $\Delta\tau$ leads to an increase in sensitivity and hence a superlinear dependence of photocurrent on excitation intensity.

(3) $ef\tau \Delta\mu_n$ relates to mobility changes that result from an increase in the density of ionized impurity scattering centres, inhomogeneities of crystal potential or transitions from lower to higher mobility states.

The sensitivity S of a photoconductor is defined quantitatively by the expression[2]

$$S = \frac{(\Delta i/V) l^2}{P} \tag{5.4}$$

where Δi is the photocurrent, V is the applied voltage, l is the electrode spacing and P is the absorbed radiant power. The best photoconductors have an S value of the order of unity; S is also directly proportional to τ and μ_n.

A more familiar quantity used to express the performance of a photoconductor is the photoconductive gain, G, which for a single carrier type is given by

$$G = \frac{\tau}{T_R}$$

where T_R is the transit time between electrodes

Therefore

$$G = \frac{\tau}{(l^2/\mu_n V)} = \frac{\tau \mu_n V}{l^2} \tag{5.6}$$

thus

$$G \propto \frac{V}{l^2} S \tag{5.7}$$

Needless to say the definitions given above to describe photosensitive behaviour are not unique and often other definitions such as light to dark current ratio may be more convenient.

The term "intrinsic" is applied to the photoconductive behaviour of the purest obtainable crystals of the II–VI compounds although they do contain many natural defects which are in effect impurity states. If one accepts the description "intrinsic" behaviour with suitable reservations, it is fair to say that such behaviour is not characterized by high photosensitivities. This is particularly so in large energy gap

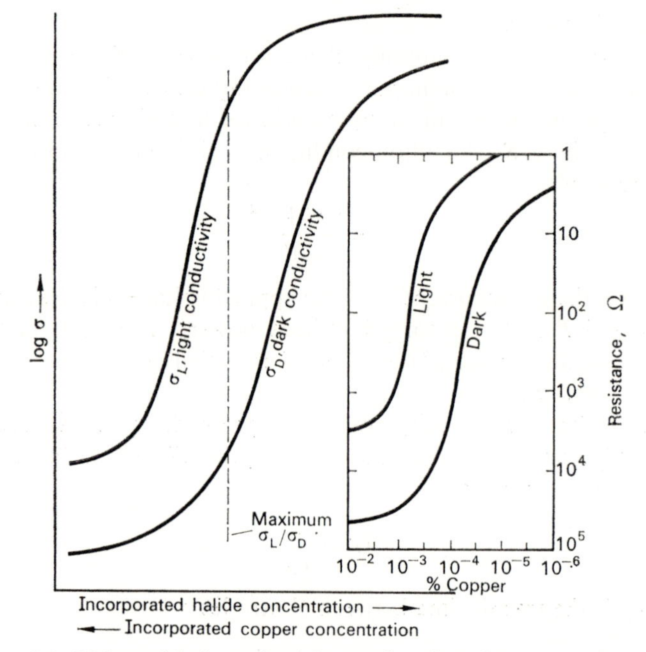

FIG. 5.1. Light and dark conductivity as a function of the copper impurity concentration.[1]

materials where recombination effects occur very rapidly and carriers are less mobile. Enhancement of the photosensitive behaviour of a material is achieved by the deliberate introduction of impurity elements as is evidenced by the studies of Bube on CdS.[2, 3] The introduction of either donor or acceptor impurities into CdS can lead to highly photosensitive material. Iodine is a donor, which gives conducting

CdS, and generates acceptor-like cation vacancies (V_{Cd}) in an attempt to achieve charge compensation; the cation vacancies take on a negative charge with respect to the cation sublattice and accordingly have high capture cross-sections for photoexcited holes and low capture cross sections for electrons. The introduction of copper as an acceptor provides insulating CdS which is also highly photosensitive as a result of the negatively charged copper sensitizing centre on the cation sublattice. The effect of these deliberately added sensitizing centres

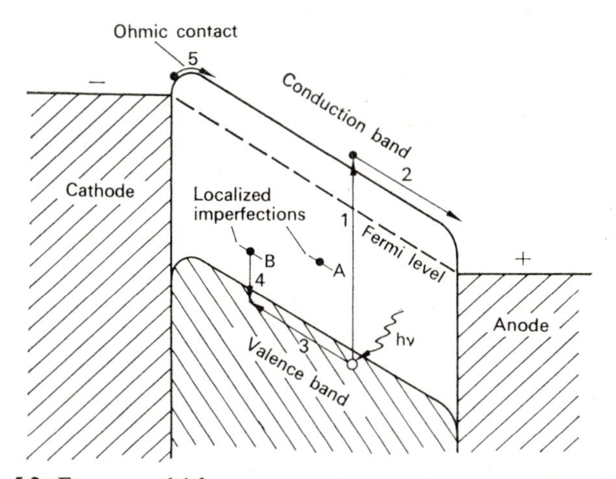

FIG. 5.2. Energy model for an n-type photoconductor in operation.[2]

is to increase the free electron lifetime from 1 μs to 10 ms in spectrographically pure CdS. There exists an optimum concentration of the halogen or copper for which the ratio of light to dark conductivity is a maximum; Fig. 5.1 illustrates the dependence of the light current under fixed illumination conditions and the dark current on the copper concentration.[1]

An energy model has been proposed for the sensitization of an n-type photoconductor on the basis of two competing centres A and B and is shown in Fig. 5.2.[2] The centres, A, have a high probability of capturing a free electron after initial capture of a hole, whereas the

B centres have only a small probability of free electron capture in similar circumstances. The deliberate introduction of the B centres in the preparation of the material results in a sensitive photoconductor.

5.1.2. *Spectral Distribution and Speed of Response*

The photoconductivity as a function of the photon energy clearly depends on the ability of the material to absorb the incident radiation. A peak in the response occurs approximately at the energy for which the absorption constant is equal to the reciprocal of the crystal thickness. At higher energies the absorption constant is much larger with the result that only surface regions are excited where defect states give shorter lifetimes and a consequent drop in photosensitivity. On the other hand radiation of longer wavelength is only partially absorbed and the resulting photocurrent is less than the maximum. The presence of defect centres extends the spectral response to longer wavelengths because of direct excitation of carriers from defect levels into the conduction band.

The speed of response of a photoconductor is dependent upon the centres which affect the free carrier lifetime. If recombination centres predominate then the response time is determined by the free carrier lifetime. However, at low excitation intensities the speed of response is orders of magnitude slower because the initial small density of photoexcited carriers is rapidly trapped; only after a steady state is established is the maximum photocurrent reached. This trapping effect also lengthens the decay time since electrons are slowly released after removal of the excitation source. Response times greater by between 10^3 and 10^6 times than the free carrier lifetime have been observed at low excitation intensities. There is one partially incompatible feature in the combination of rapid response time and high photosensitivity at low excitation levels in that the speed of response requires short free carrier lifetimes, the reverse of what is required for high photosensitivity, or alternatively the total absence of electron traps a somewhat utopian wish in II–VI compounds.

An improvement in response time for low excitation intensities would greatly benefit the practical usefulness of photoconductors. Two approaches to the subject have been suggested:[(2)] (1) decrease the density of traps that influence the behaviour of the free carriers; (2) adjust the sensitizing centre energy level such that the free carrier lifetime decreases during decay and is also shorter during initial excitation. The first way requires strict control over the preparation conditions and for really low trap densities single crystals are needed. From an applications point of view single crystals have limited usefulness and as such this approach is more of academic interest than anything else. The second possibility for increasing the speed of response is more promising and requires the sensitizing centre to operate just beyond the hole trap–recombination centre boundary. A decrease in light intensity causes the centre to behave as a hole trap, thus on removal of excitation both electrons and holes will be trap released to give a reduced decay time. The ionization energy E for holes from sensitizing centres is fixed for a given photoconductor and to successfully adjust the location of this energy level alloys between different compounds have been suggested. The fact that $E = 0{\cdot}6$ eV for CdSe and $1{\cdot}1$ eV for CdS indicates why at room temperature CdSe has a faster decay than CdS. An alternative approach to the use of alloys is to introduce a high density of sensitizing defects but here one would have to suffer a considerable decrease in photosensitivity.

5.1.3. *Thermal and Optical Quenching of Photoconductivity*

Quenching occurs when the thermal and optical energies supplied to the photoconductors are sufficient to excite electrons from the valence band into sensitizing centres and convert them from recombination centres into hole traps. Under these conditions the photosensitivity decreases and the material becomes less useful as a photoconductor.

(a) *Thermal Quenching*. The effect of thermal quenching is most easily understood if the dependence of photoconductivity on light intensity and temperature is observed. Figure 5.3 illustrates the point with a plot of photocurrent as a function of temperature for a range of excitation intensities in CdSe:I, Cu crystals.[8] The implications of these curves are that the excitation level and the temperature have complementary effects in modifying the behaviour of the sensitizing centre. A model similar to that mentioned in section 5.1.1 is used to describe this behaviour;[9] it assumes the existence of two different kinds of recombination centres and also electron trapping states located energetically close to the bottom of the conduction band. Centre A has a large capture cross-section for electrons, whereas centre B, which is deliberately created to sensitize the photoconductor, has under appropriate excitation conditions a low capture cross-section for electrons; the traps, as already mentioned in section 5.1.2, give rise to a slow response at low excitation levels. As long as centre B remains

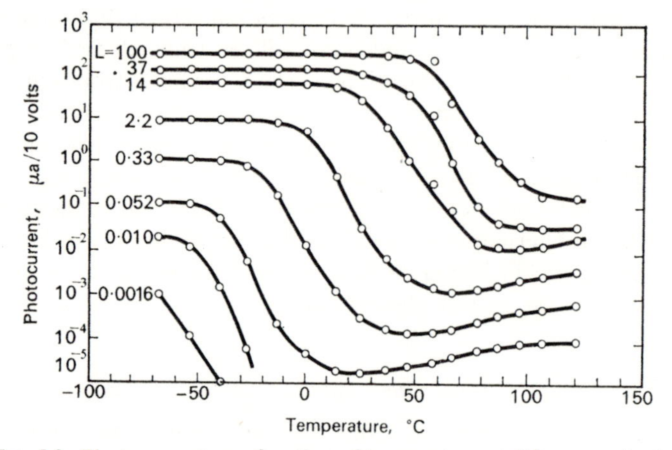

FIG. 5.3. Photocurrent as a function of temperature at different excitation intensities in CdSe:I, Cu crystals.[8]

in thermal equilibrium with the valence band, the behaviour of the photoconductor is determined by centre A. However, when hole capture by centre B becomes the dominant process, the electron population of the centre is effectively transferred to centre A. The result is that centre A is rendered inactive as a recombination centre and the photoconductive behaviour is determined by the major property of centre B, namely a low electron capture cross-section. The change in the character of centre B from that determined by thermal equilibrium with the valence band to that determined by recombination kinetics leads to an increase in the photosensitivity; hence in the transition process from dominance by A centres to that by B centres the photocurrent shows a superlinear dependence on excitation intensity. If one refers to equation (5.4), this represents the condition under which a positive change in free carrier lifetime $\Delta\tau$ occurs; hence assuming $\Delta\mu_n$ changes are small, the conductivity has the lifetime effect superimposed on the linear excitation intensity effect and superlinear dependence of conductivity on excitation intensity results.

With an increase in temperature the hole occupancy of centre B starts to change and an approach towards thermal equilibrium can occur. Such an effect results in a fall in photosensitivity and a sublinear dependence of photoconductivity on excitation intensity comes into operation. Again equation (5.4) may be used to justify this statement in that $\Delta\tau$ negative will produce an effective sublinear dependence on excitation intensity. In Fig. 5.3 the decrease in temperature, at which thermal quenching effects become important with decreasing intensity, is simply explained by the smaller number of holes available to cause reversion of the centres B from thermal equilibrium behaviour with the valence band to recombination centre characteristics. It may also be noted in this figure, that at higher temperatures and the low excitation intensities particularly, a rise in the desensitized photocurrent results from the thermal release of electrons from the trapping states close to the conduction band. Figure 5.4 illustrates for the CdSe:I, Cu crystal the photocurrent as a function of excitation intensity plotted on logarithmic axes at a range of temperatures.[8] At high excitation intensities, near linear behaviour is observed for low and intermediate temperatures and is satisfactorily explained by the complete conversion of the centres B from hole traps into recombination centres at relatively low excitation intensities. The photoconductivity behaviour at these high excitation levels will be linear provided that the carrier density is not sufficient to cause a decrease in mobility or affect the lifetime, i.e. $\Delta\mu_n$ and $\Delta\tau$ must equal to zero in equation (5.4). The range of photocurrents over which superlinear behaviour is found is of the order of 10^6.

It is possible with an analysis of the kinetics of the different centres to derive quantitative information about the location of the sensitizing centre and the ratio of the hole and electron capture cross-sections of the centre. It is the transitions from near linear or sublinear to superlinear behaviour when considered as a function of temperature which furnish the required information. The slope of the graph of the logarithm of the majority carrier density versus $1/T$, for both the high and low excitation transition currents, is used to determine the hole

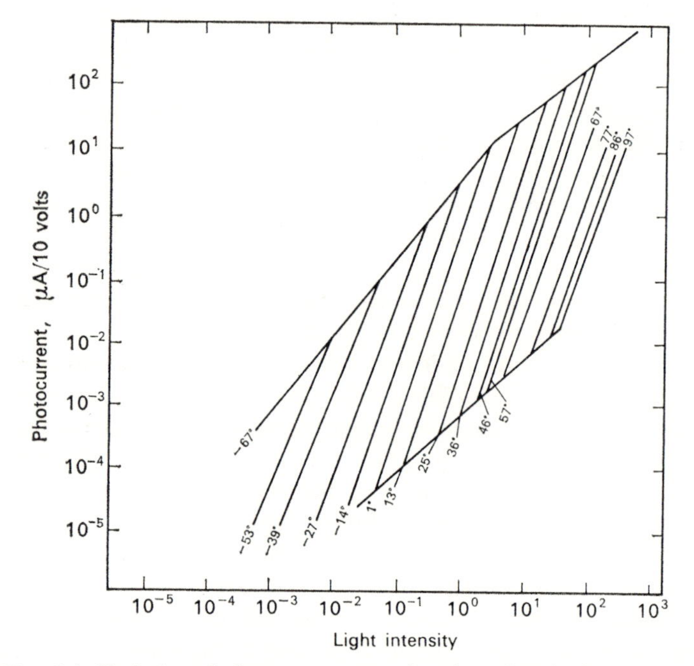

FIG. 5.4. Variation of photocurrent as a function of excitation intensity at different temperatures for CdSe:I, Cu crystals.[8]

ionization energy of the sensitizing centre. The intercept of the line with the $1/T$ axis for the high excitation intensity transition provides the ratio of the hole to electron capture cross-sections of the sensitizing centres. Table 5.1 summarizes the results obtained by Bube[10] from such an analysis on a range of compounds, which include CdS, CdSe, ZnSe.

The range of values obtained for the capture cross-section ratios and the hole ionization energies of the II–VI compounds is a result of the large number of crystals investigated under a variety of doping conditions.

TABLE 5.1. CROSS-SECTION RATIOS AND IONIZATION ENERGIES OF SENSITIZING CENTRES IN SEVERAL COMPOUNDS[10]

Compound	Energy gap (eV)	Hole-ionization energy (eV)	Capture cross-section ratio
CdS	2·4	1·2	2×10^6–5×10^8
CdSe	1·7	0·6	2×10^3–10^8
ZnS	3·7	1·2	
ZnSe	2·6	0·6	5×10^2–6×10^8
GaAs	1·4	0·4	6×10^6
GaP	2·2	0·7	
InP	1·2	0·4	7×10^4
ZnTe	2·1	0·3*	
GaSe	2·0	0·5*	4×10^4

* There is some indication that the photoconductivity in ZnTe and in GaSe is *p*-type, in which case the quoted values are for electron-ionization energies.

(b) *Optical quenching.* Optical quenching can be used as an alternative approach to thermal quenching in the derivation of information about the sensitizing centres. The optical frequency required to observe optical quenching effects in the II–VI compounds is generally in the near or medium infrared. In some compounds the infrared radiation, which is used to quench the conductivity induced by band gap photons, can separately create its own photocurrent as instanced by the work of Ullman and Dropkin,[11] Kang[12] and Blount[13] on ZnS:Cu crystals. The infrared photocurrent in ZnS showed a similar spectral dependence to that of the infrared quenching. An enhancement of the 3650 Å U.V. excited photocurrent was observed in some crystals when they were subjected to infrared irradiation; the enhancement of photocurrent was by as much as 50% for radiation of wavelength 0.7 μm. However, the enhanced photocurrent was always less than the sum of separate ultraviolet and infrared photocurrents, and in effect, there was a limited amount of optical quenching. It was concluded that the infrared

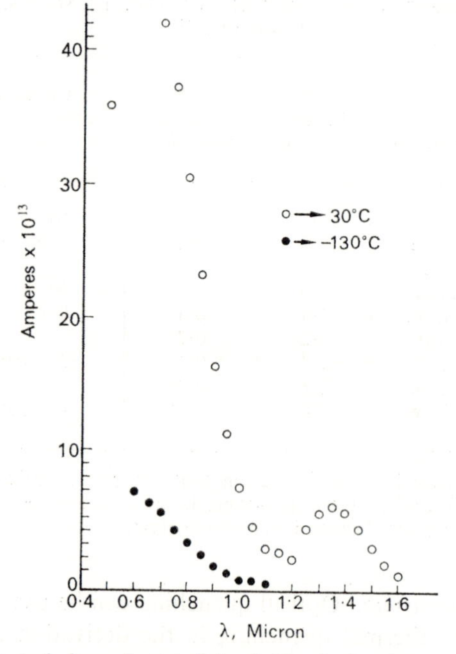

FIG. 5.5. Spectral dependence of the infrared photocurrent in ZnS:Cu at − 130°C and 30°C.[11]

photocurrent resulted from hole flow in the valence band of the zinc sulphide. Quenching of the U.V. excited photocurrent by the infrared radiation occurred in many of the ZnS crystals and again the spectral dependence was similar to that of the infrared photocurrent. Figure 5.5 shows the spectral dependence of the infrared photocurrent in ZnS:Cu at − 130° and 30°C.[11] It is seen that at low temperatures the long wavelength infrared peak disappears and explanations for this are usually based on the removal of the possibility of thermal excitation of holes from an excited state of the impurity centre into the valence band.

The optical quenching of photoconductivity in CdS has been studied over a wide range of photosensitivities and has revealed a general pattern for the phenomenon. Figure 5.6 illustrates three examples of

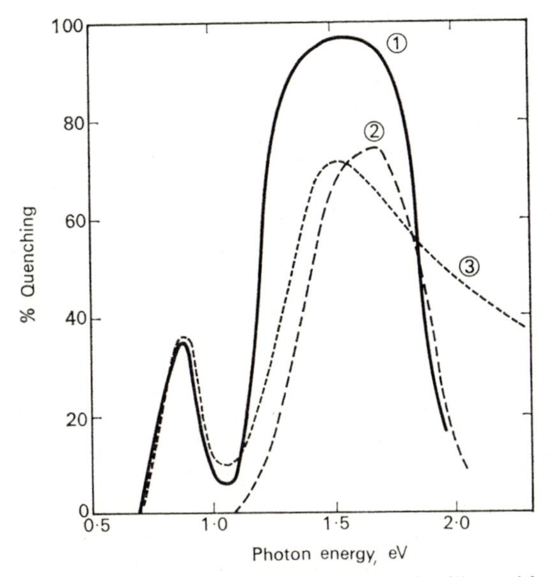

FIG. 5.6. Optical quenching spectra for CdS crystals: (1) sensitive crystal, (2) insensitive crystal, (3) very pure crystal.[1, 3]

the different forms of optical quenching spectra observed in CdS crystals: (1) The sensitive crystal has an extremely sharp rise in quenching as the photon energy increases from a threshold at 1·1 eV; quenching almost reaching 100% at 1·5 eV and then falls away to zero by 2·0 eV. A low energy quenching peak at 0·85 eV also occurs; (2) The insensitive crystal has a slower rise in quenching from a threshold at 1·1 eV but again falls away to zero at around 2·0 eV;

(3) The quenching behaviour of the pure crystal is similar to that of the insensitive crystal except at higher energies, where appreciable quenching is observed to within 0·1 eV of the absorption edge.[14] The general conclusions that may be drawn from these characteristic quenching spectra are that, (a) the onset of quenching is most marked in the sensitive crystals, although appreciable quenching is observed in other less sensitive crystals and (b) the quenching continues almost to the absorption edge but is masked in doped crystals by simultaneous excitation from the defect centres. The low energy quenching peak at 0·85 eV is similar to that observed in ZnS:Cu[11] and results from thermal excitation of holes from the excited state of the defect centre into the valence band. Again low temperature investigations indicate the disappearance of the 0·85 eV peak below 200°K.[1]

Analysis of the optical quenching effects of the photocurrent as a function of the intensity of the infrared radiation permits a determination of not only the capture cross-section ratio but also the electron capture cross-section of the sensitizing centre. The photocurrent at the transition point from superlinear to sublinear behaviour is used to determine the electron capture cross-section of the sensitizing centre. The photocurrent at fixed excitation and infrared quenching intensity can also be used to determine the capture cross-section as a function of temperature. An electron capture cross-section of the sensitizing centre in highly photosensitive CdS equal to 10^{-21} cm^2 was obtained and the temperature dependence was found to be very slight.[3] This slight temperature dependence of the capture cross-section suggests a neutral centre, such as electrically neutral associated cation vacancy and donor impurity as the sensitizer; spin resonance studies appear to confirm this picture of the centre. Comparison of the optical ionization energy of the centre with the difference between the band gap energy and the peak energy of the self-activated luminescence, lends further support to the idea of an associated cation vacancy and donor impurity as the sensitizing centre in ZnS, CdS and ZnSe. Table 5.2 summarizes the hole ionization energies derived from optical quenching data for ZnS, CdS, ZnSe, CdSe.

The general pattern appears to be that the ionization energy E_I of the sensitizing centre derived from both thermal and optical quenching experiments decreases in magnitude over the sequence sulphide, selenide, telluride, E_I is 1·1 eV for the sulphides, 0·6–0·7 eV for the selenides, and 0·3–0·4 eV for the tellurides. Such behaviour is consistent with sensitizing centres formed by defects on the cation sublattice as suggested in the preceding discussion.

TABLE 5.2

Compound	Optical quenching peaks (eV)	Optical ionization energy of the sensitizing centre (eV)
ZnS:Cu, Cl[11, 13, 19]	0·97, 1·5	1·2
CdS[1, 15, 17, 19]	0·85, 1·5	1·1
CdS:P[18]	0·85, 1·5, 2·15	1·1, 2·15
ZnSe[3]		0·66–0·78
ZnSe:Cu, Br[16]	1·4	0·6
ZnSe:Sb, Br[16]	1·4, 1·6	0·7, 1·2
CdSe[15, 17]	1·1	0·6–0·68

5.1.4. *Trapping States*

The presence of electron trapping states in II–VI compounds always sets a limit to the performance of the material as a photoconductor. Trapping effects at low excitation intensities lead to a slower response time and at all levels of excitation lead to a long decay-time. Further, if the electron trapping levels are situated close to the Fermi level of the excited photoconductor, then a considerable desensitization will exist until the traps are filled. Hence, it can be seen that in the assessment of the suitability of material as a satisfactory photoconductor knowledge of the trap density and energy location are very necessary. Over the years many analyses have been made of properties which result from electron trapping and the classification of materials produced under different growth conditions is reasonably well documented. The principal phenomena that have been used to study trapping states are photoconductive rise and decay, thermally stimulated conductivity glow and space charge limited currents. The measurements of these properties in the determination of trapping states in photoconductors is usually made on crystalline samples with two electrical contacts across them and a D.C. potential applied. Mention was made of the conductivity glow characteristic in section 4.2 as an alternative method in the investigation of trapping states to thermoluminescent emission in compounds with appreciable conductivities. The basic ideas behind these two phenomena are the same and therefore the description given below will assume the theory discussed in section 4.2.

(a) *Photoconductive rise and decay.* It is possible with a thermally cleaned material to derive information about the density of trapped carriers from the initial photoexcited current rise. A proportion of the optically excited electrons will be trapped and provided that no electrons are released from the traps while the photocurrent rises to its steady value, the area between the growth curve and the equilibrium value of photocurrent can be directly related to the trapped carrier density. The photoconductive decay is proportional to the rate of release of electrons from traps and takes an exponential form for a single trap energy with no retrapping. The presence of several trapping states, as is generally the case in II–VI photoconductors, leads to a complex form of photoconductive decay, which is difficult to interpret.

It has been shown by different investigators[10, 20] that it is possible to analyse trapping states in CdS from regions of exponential photocurrent decay. Nicholas and Woods[20] have divided the decay process into three distinct parts: (1) initially fast decay associated with free electron recombination occurs, (2) then an intermediate region exists where electrons start to be released from traps and (3) finally, decay associated with emptying of trapping levels close to the equilibrium Fermi energy occurs. The latter part of the decay, which is exponential in form, is described by

$$\sigma = n_{to}\mu e\tau p \exp(-pt) \qquad (5.8)$$

where n_{to} is the number of traps filled at temperature T_o, μ is the carrier mobility, e is the unit of electronic charge, τ is the free carrier lifetime and p is the electron escape probability from a trap at depth E given by

$$p = N_c v S \exp\left(-\frac{E}{kT}\right) \tag{5.9}$$

N_c is the effective density of electron states in the conduction band, v is the electron's thermal velocity and S is the electron capture cross-section of the trap. Measurements of the decay constant, $\tau_d = 1/p$, over a range of temperatures permit both the determination of the trap depth and the electron capture cross-section from the plot of $\log_e \tau_d$ against $1/T$. If retrapping of electrons occurs, then the decay of photoconductivity follows a power law dependence and the interpretation requires a more complex analysis.

Bube *et al.*[21] have analysed the example of a quasicontinuous trap distribution which is assumed in thermal equilibrium with the conduction band during the period of trap emptying. Consequently the quasi-Fermi level E_{fn} will provide an indication of the trap depth. The density of free electrons n is given by

$$n = N_c \exp\left(-\frac{E_{fn}}{kT}\right) \tag{5.10}$$

The decay time τ_o is given by the time required for the number of free electrons to fall to $1/e$ of the initial value and in this time the quasi-Fermi level changes by an energy kT. Thus it is possible to calculate the density of electron traps n'_t in the energy interval kT, which is

$$n'_t = n\tau_o/\tau \tag{5.11}$$

and the trap depth which is effectively E_{fn} is found from equation (5.10). Figure 5.7 illustrates the trap density as a function of energy as determined by a Fermi level analysis of decay curves measured at 18 different temperatures between 90° and 370°K for a $CdS_{0\cdot73}Se_{0\cdot27}$ crystal.[21]

Bube[3] has noted the difficulties in distinguishing between the existence of several discrete trap levels without retrapping, several discrete trap levels with retrapping and a quasi-continuous trap distribution from photoconductive decay observations. The need for independent support of the results obtained from any single method is important.

(b) *Thermally stimulated conductivity.* Thermally stimulated conductivity occurs when samples that have been subjected to excitation by band gap radiation at low temperatures are heated. Bursts of current occur at temperatures which characterize trapping states being thermally

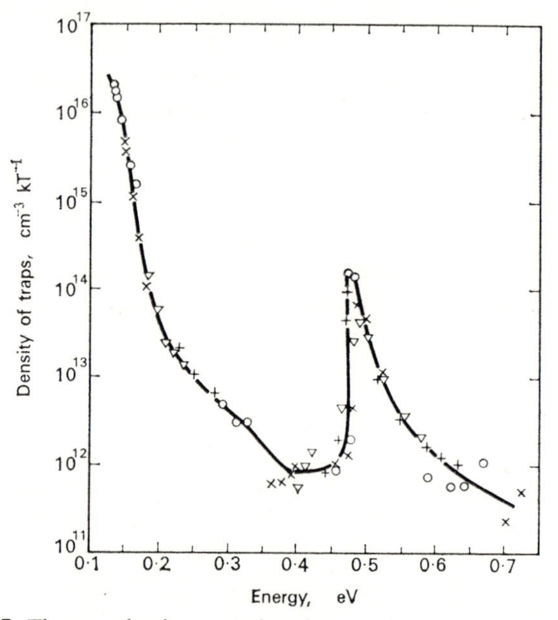

FIG. 5.7. The trap density per unit volume and per kT as a function of trap depth as determined by a Fermi-level analysis at temperatures between 90 and 370°K.[21]

emptied. The various methods for the interpretation of thermally stimulated conductivity (TSC) curves have been the subject of several review articles.[20, 22] The conclusion that may be drawn from these reviews is that unambiguous interpretation of the nature of trapping states is difficult. Most of the methods assume extreme conditions such as slow or fast retrapping and therefore are not always applicable.

Bube *et al.*[21] have made some interesting TSC studies in conjunction with photoconductive decay measurements described above on CdS–CdSe crystals. They assumed fast retrapping and determined the trap depths from TSC glow peaks on the assumption that traps were half emptied at the peak and hence the Fermi energy was equal to the trap depth. A distribution in trap density with energy very similar to that derived from photoconductive decay was observed and is shown in Fig. 5.7. A relatively high density of traps at 0·20 and 0·45 eV was

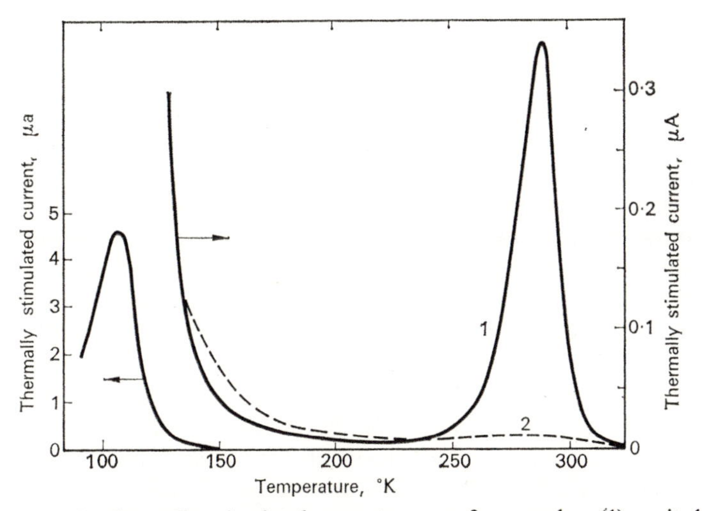

FIG. 5.8. Thermally stimulated current curves for samples, (1) excited continuously while cooling and (2) excited at 90°K only.[21]

found in these investigations. A glow peak at 289°K in the same crystals could only be observed if the crystal was excited during cooling from room temperature. Figure 5.8 illustrates the TSC glow curves for the sample when it is (1) excited during cooling and (2) excited only at 90°K. Further analysis of this trap from the slope of decayed TSC curves indicates two activation energies for it of 0·43 eV, and 0·75–0·85 eV, Fig. 4.11. The trap is thought to be located 0·43 eV below the conduction band and is surrounded by a repulsive barrier of 0·30 eV height; the electrons can escape either by tunneling through the barrier or passing over the full barrier height, hence the two activation energies. A trap similar in character to the above has been found in CdS by Woods and Nicholas,[23] which they ascribed to photochemical effects.

Cowell and Woods[24] have used the technique of curve fitting in the evaluation of TSC curves for CdS crystals. They have analysed curves both for slow and fast retrapping conditions and have used a system of successive approximations to obtain the best curve fit. Dussel and Bube[25] have developed expressions to describe the TSC behaviour for a single shallow trap depth in the presence of deeper traps and a single type of recombination centre without restriction on the kinetics. However, application of this analysis to determine the trapping parameter from the TSC curves appears to be complex.

In an extensive analysis of the rate processes which determine the behaviour of electrons, Haine and Carley-Read[26] have arrived at three general methods for the determination of trapping parameters. The only approximation made in the analysis is that dn_c/dt, the rate of change of free electron density with time, can be neglected and no assertions have to be made as to the retrapping. Trap depths, E_t, had been derived from (a) the slope at any point on the TSC curve, (b) the TSC peak expression and (c) an integrated area up to any point on the curve related to the recombination of free carriers; E_t values determined for a given CdS crystal by the three methods had a spread of little over 5% and were in agreement with values found from Bube's Fermi level analysis; the results thus suggest that fast retrapping mechanisms are present in the CdS crystals studied. They concluded that the slope method had the advantage over other methods in that a point on the TSC curve could be chosen where there was no interference from other peaks. Other trapping parameters which have been determined by their analysis are N_t, the trap density, n_{to}, the initial trap occupancy, and S_t, the trapping cross-section, with τ, the recombination lifetime, and τ_t, the retrapping time for electrons, determined independently. The same authors have devised an entirely new approach to the analysis of TSC curves with a constant temperature study.[26] In the constant temperature method the crystal is heated until a particular trapping level starts to empty, then the temperature rise is stopped;

with the temperature held constant the current decay as a function of time is studied and correlated with the relevant theory. A plot of $\log_e n_c$ versus time indicates which kind of trapping is present, if the trapping states had been initially filled; for fast retrapping the plot gives two different linear slopes, θ/τ_t followed by θ/τ, and for slow retrapping it has only one slope θ/τ, where θ is the trapping factor for the level. Absolute values of E_t, S_t, N_t, τ_t and θ have all been determined from the constant temperature analysis.

(c) *Space-charge limited (SCL) current trap determination.* Large deviations from the space-charge limited current values derived for a trap free insulated case suggested that the analysis must take account of electron trapping states. Lampert[27] developed a theory of single-carrier SCL conduction in an insulator with traps. The theory provides a current–voltage characteristic with several distinct regions. Ohm's law is obeyed at low voltages when the injection of excess carriers at the cathode is negligible and the true resistivity of the crystal is obtained, $\rho_o = 1/n_o e\mu_n$. At a voltage V_1 the current starts to increase more rapidly with voltage and in the absence of traps this would occur at

$$V_1 = \frac{ed^2}{2\kappa\varepsilon_o} n_o \tag{5.12}$$

where d is the electrode spacing, n_o is the thermal equilibrium concentration of electrons in the conduction band, the other symbols having the same meaning as given at other points in the text. However, in the insulator some of the injected charge becomes localized in traps where, although it does not contribute to the current, it forms part of the space charge. For a single trap depth E_t of density N_t, it can be shown that the current density J is given by

$$J = \frac{9}{8}\frac{\kappa\varepsilon_o\mu_n V^2}{d^3}\theta \tag{5.13}$$

where θ, the trapping factor, is equal to the ratio of the free electron density n_c to the trapped electron density n_t

$$\theta = \frac{n_c}{n_t} = \frac{N_c}{N_t}\exp\left(-\frac{E_t}{kT}\right) \tag{5.14}$$

Since most of the injected charge is trapped, then the expression for the transition voltage, equation (5.12), becomes modified to

$$V_1 = \frac{ed^2}{2\kappa\varepsilon_0\theta} n_0 \tag{5.15}$$

Finally at some still higher voltage, V_2, according to Lampert's theory, all the available traps become filled and the current increases discontinuously by $1/\theta$ to the theoretical SCL current in a trap-free insulator. Lampert has shown that at this limit

$$V_2 = \frac{ed^2}{2\kappa\varepsilon_0} N_t \tag{5.16}$$

Thus in theory both N_t and E_t can be obtained from a space-charge limited current analysis.

Marlor and Woods[28] have analysed the SCL current behaviour in CdS crystals in order to determine the density of traps and the trap depth. They have derived values for N_t by independent methods and the value given by equation (5.16) is an order of magnitude lower than that obtained from a free electron method. This fact coupled with similar evidence from other investigators lead them to suggest that V_2 is not a trap-filled limiting voltage and any values of N_t derived from equation (5.16) should be treated with caution. Tredgold,[29] however, has pointed out that the theory used does not include a carrier diffusion term which could increase the N_t value obtained by roughly an order of magnitude and therefore overcome the inconsistency in N_t values. The trapping parameters found from the SCL current behaviour in CdS crystals were $N_t = 10^{14}\ \text{cm}^{-3}$ and $E_t = 0{\cdot}61$ eV; it is assumed that the trap at 0·61 eV predominates over all the other traps in determining the form of the current–voltage characteristic.[28, 30] Space-charge limited current behaviour, if studied at a range of temperatures, should provide information about the electron trap which dominates the current level. In practice the SCL current analysis has only been used to any extent at room temperature. TSC behaviour and photoconductive decay have provided most of the information about trapping states in the II–VI photoconductors.

Table 5.3 summarizes the trap depths derived by the different analyses of TSC, photoconductive decay and SCL current characteristics in a CdS crystal after Haine and Carley-Read.[26]

TABLE 5.3. TRAP DEPTHS (eV) FOR A CdS CRYSTAL DERIVED FROM DIFFERENT ANALYSES[26]

Analysis	E_{t2} (eV)	E_{t_1} (eV)
Slope expression	0·26	0·56
Peak expression	0·25	0·54
Integral expression	0·26	0·52
Bube Fermi level analysis	0·27	0·54
Initial exponential rise	0·08	0·27
Peak expression (monomolecular kinetics)	0·08	0·40
Half peak temperature and area	0·05	0·20
Peak shift as a function of heating rate	0·04	0·28
SCL current behaviour	—	0·61
Photoconductive decay	0·05	0·24

5.2. Particular Photoconductive Characteristics of the II–VI Compounds

5.2.1. *Cadmium Sulphide*

Bube[3] has drawn attention to the importance of traps in determining the speed of response of CdS crystals. Crystals grown from the vapour phase were subjected to extreme conditions of control in relation to the purity of the carrier gas, the flow rate and the temperature. The trap density was reduced from approximately 10^{15} cm^{-3} to 5×10^{13} cm^{-3} and resulted in a 500 times faster response time. The introduction of a trace of iodine is particularly important in the speed improvement, in that it creates shallow traps at the expense of the deeper traps resulting from lattice defects. Figure 5.9 illustrates the TSC curves for CdS crystals in which the trap density is progressively reduced and response speed consequently improved. In the slowly responding crystal the room temperature TSC peak is dominant, whereas in the fastest crystal obtained the 100°K peak is beginning to dominate the trapping process.

Nicholas and Woods[20] in their analysis of trapping states identified as many as six different discrete sets of traps in CdS crystals. The trap

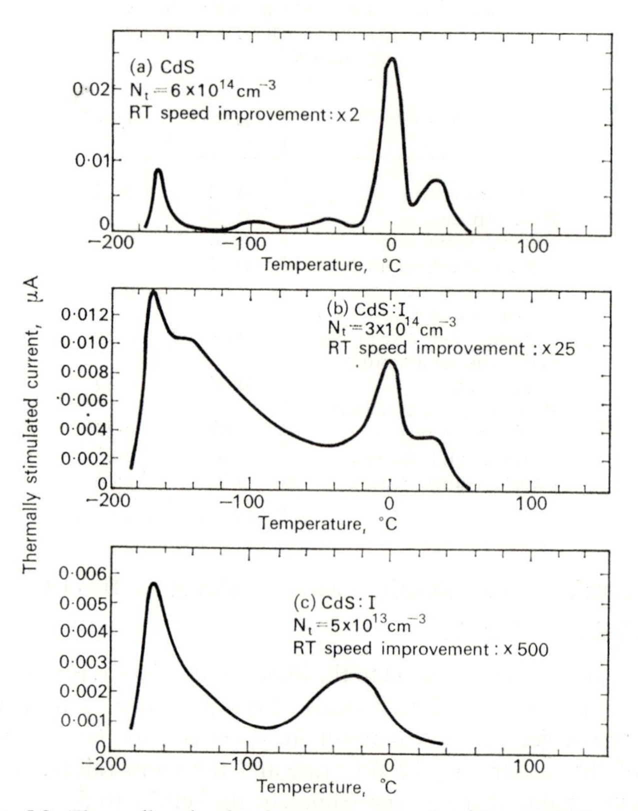

FIG. 5.9. Thermally stimulated current curves for (a) pure CdS crystal, (b) and (c) pure CdS crystals with a trace of iodine[3] (*RT* refers to room temperature).

depth values were obtained using the technique of thermal cleaning in which a given peak is isolated from its surrounding peaks by a series of decayed glow curves; trap depths of 0·05 eV, 0·14 eV, 0·25 eV,

0·83 eV and 0·63 eV were obtained with this form of analysis. The first three values for the trap depths given above differ greatly from the Fermi level interpretation in which E_t would be 0·20, 0·32 and 0·52 eV. Skarman[31] has correlated the trapping states observed in single crystals and polycrystalline thin films of CdS from both TSC curves and photoconductive decay. Crandall[32] has made some interesting observations of trapping distributions in CdS single crystals between 4 and 70°K from the temperature dependence of photocurrent. He found an exponential increase of photocurrent with temperature for undoped and copper-doped insulating crystals 0·1–1 mm in thickness. The crystals when excited by 465 mμm radiation with absorption in the first 0·1 mμm of the surface the exponential behaviour occurred between 4·2 and 30°K, whereas with uniformly absorbed 700 mμm radiation the exponential form did not go below 15°K. It has been possible to fit this behaviour to an exponential distribution of traps of the form

$$N_t(E) = A \exp(-E_t/kT_1) \qquad (5.17)$$

where T_1 is a temperature characteristic of the distribution and values obtained are, for 465 mμm excited undoped crystal $T_1 = 46°$K, for 465 mμm excited Cu-doped crystal $T_1 = 64°$K and for 700 mμm excited Cu-doped crystal $T_1 = 91°$K. The fact that with the uniformly absorbed radiation the photocurrent is constant below 15°K merely indicates there is a limit to the exponential distribution at small energies below the conduction band (0·004 eV).

In an analysis of the effect of heat treatment of CdS single crystals in sulphur and cadmium atmospheres much information has been derived about the origin of the peaks in the photoconductivity spectrum near the band edge.[33–35] CdS single crystals as grown contain sulphur vacancies, cadmium vacancies and cadmium interstitials and associated complexes of these defects. Crystals annealed in an atmosphere of cadmium vapour at low pressures have the conductivity reduced by several orders of magnitude and cadmium vacancy defects are removed.[33] Uchida[35] has identified peaks in the photoconductivity spectra of CdS crystals which are likely to be associated with cadmium vacancies. He has also used excess sulphur vapour pressures in anneals at 850°C to remove sulphur vacancies and has been able to identify more fully the photoconductivity peaks. Table 5.4 illustrates the peaks in the photoconductivity spectrum in a CdS crystal and the nomenclature used to identify them.

Peaks P_2 and P_4 decreased rapidly with increase in sulphur pressure and after their disappearance, P_5 decreased. Further increases in the sulphur pressure and in the duration of the anneal removed first P_3 and then P_1. Applied cadmium vapour pressures create P_3 as the dominant peak and P_1 as a subsidiary peak; P_5 also grows with additional cadmium pressure increase. P_6 and P_7 are thought to be introduced as a result

TABLE 5.4. LOCATION OF THE PHOTOCONDUCTIVITY PEAKS NEAR THE BAND EDGE IN A CdS CRYSTAL AT 300°K[35]

Notation	λ (mμm)	E (eV)
Exciton C	485·5	2·55
Exciton B	498	2·49
Exciton A	501·5	2·47
P_1	507·5	2·44
P_2	513·5	2·41
P_3	517	2·40
P_4	522·5	2·37
P_5	532	2·33
P_6	542	2·29
P_7	555	2·23

of accidental impurities such as chlorine and oxygen. P_1, P_3, P_5, peaks have been associated with sulphur vacancies and the electron transitions for the P_3 and P_5 peaks are thought to be valence electrons excited into V_S^+ and V_S^{++} levels. P_2 and P_4 peaks have been related to cadmium interstitials in differently ionized states. From room temperature down to liquid nitrogen temperature only peaks associated with the ground state ($n = 1$) free A, B and C excitons were observed in the photoconductivity spectrum by Uchida;[35] in addition to these peaks Park and Reynolds[34] did identify an excited state B exciton peak at 77°K along with other broad peaks under different conditions of polarized exciting radiation associated with bound A ($E \perp c$) and B ($E \parallel c$) exciton complexes. Further reduction of the temperature to 4·2°K produces a more detailed and far better resolved photoconductivity spectrum and peaks occur which are associated with excited states of A, B and C excitons.[34] The temperature dependence of the

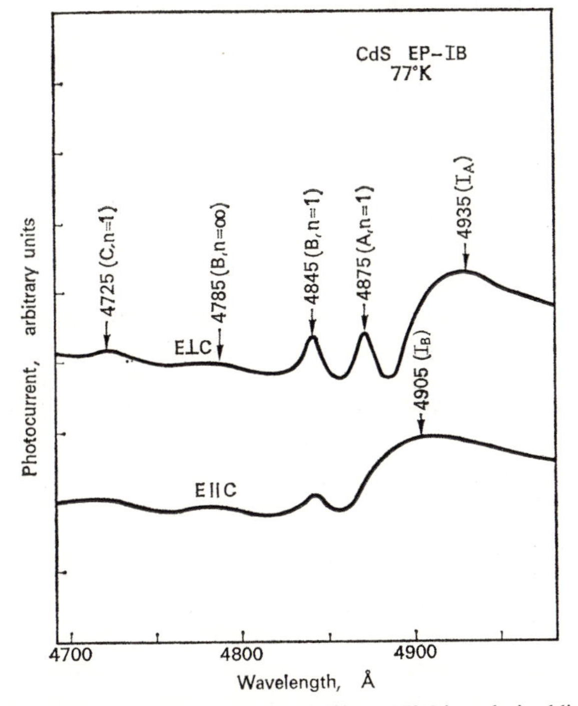

FIG. 5.10. Photoconductivity spectra of CdS at 77°K in polarized light.[35]

exciton ground state energies derived from measurements between 300 and 77°K is, as expected, similar to that for the band gap and equal to $3 \cdot 0 \times 10^{-4}$ eV/°K.[35] Figure 5.10 illustrates the photoconductivity spectra of CdS in the region of the absorption edge for radiation polarized perpendicular and parallel to the *c*-axis.[34] It is worth mentioning that if the fine structure associated with exciton and trapping states is disregarded in order to gain an overall picture of photoconductive spectral response, then the spectral sensitivity is roughly constant for wavelengths below 500 mμm and only in the ultraviolet is there an appreciable fall in sensitivity.

Park and Langer[36] observed an interesting oscillation in the photoconductivity spectrum near the absorption edge. They associated the oscillatory effect with changes in the electron lifetime, which was a minimum when the electron energy was equal to the sum of the energy of an exciton or impurity and an integral multiple of the longitudinal optical phonon energy (0·036 eV).

Photovoltaic effects have been observed from CdS in both electrochemical and *p–n* heterojunction arrangements.[38–41] The Becquerel photovoltaic effect involves the consumption of an electrode of CdS immersed in an electrolyte in order to generate a voltage.[38] The photoconductive crystal of CdS with a single indium ohmic contact forming one electrode and a calomel electrode are immersed in an aqueous solution of KCl to form a voltaic cell. Strong illumination of the CdS crystal results in a photovoltage of 600 mV being generated across the cell or a short circuit current of 100 μA. Williams[38] gives a detailed analysis of the cell reaction which can be simply described as a Cd^{2+} ion from the CdS crystal going into solution with a negatively charged S^{2-} ion left behind at the surface of the CdS electrode; removal of the two excess electrons completes the electrode reaction. The spectral response of such a cell is almost linear for wavelengths shorter than the absorption edge, i.e. $\lambda < 500$ mμm.

Photovoltages have been observed from copper diffused CdS crystals; early investigators suggested that a *p*-type CdS layer was formed on the *n*-type base CdS crystal to give a *p–n* homojunction, but the model now favoured is a conducting layer of *p*-type Cu_2S on the *n*-type CdS crystal to form a *p–n* heterojunction.[39, 40] Selle *et al.*[41] in an investigation of a Cu_2S–CdS, *p–n* heterojunction, in which a 1 μm thick layer of Cu_2S was evaporated onto the CdS crystal, observed an almost constant photovoltaic response between 500–800 mμm. Such a result tends to suggest that there is a considerable extrinsic contribution to the photovoltaic effect due to the photoabsorption at

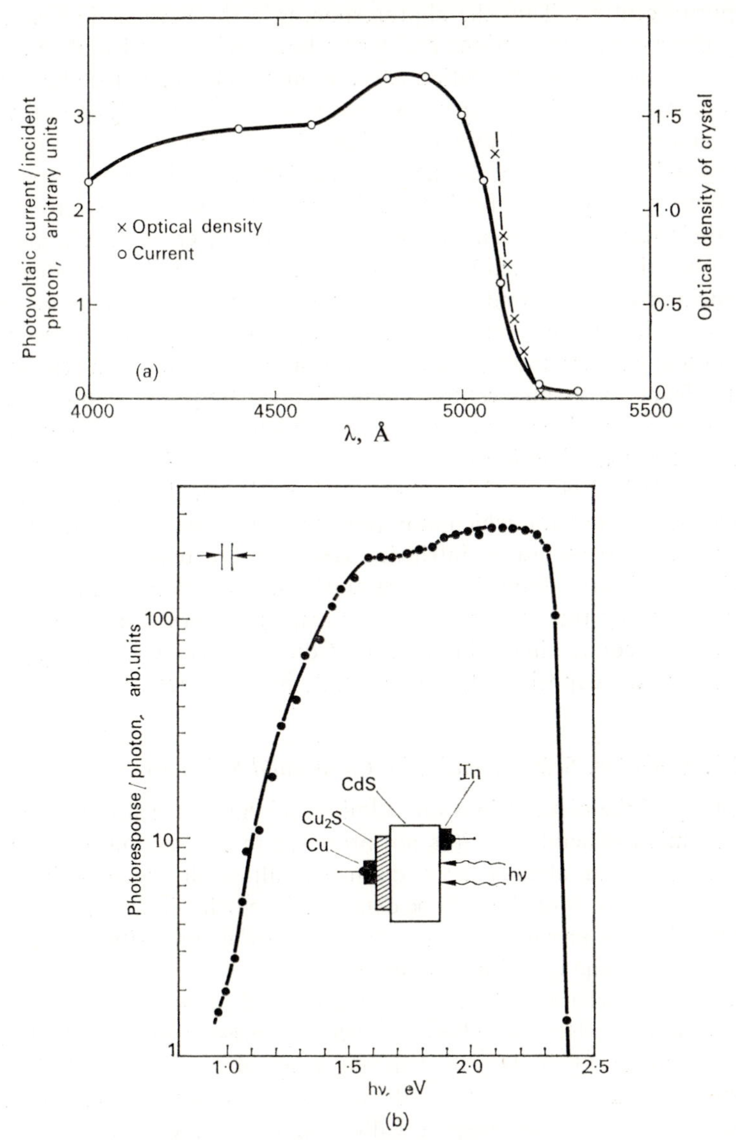

FIG. 5.11. Spectral sensitivity of the photovoltaic effect in, (a) Becquerel effect cell,[38] (b) p–n heterojunction.[41]

impurity centres. Thus the photoresponses in the Becquerel effect cell and the *p–n* heterojunctions are nearly linear with wavelength but on different sides of the absorption edge. Figure 5.11 (a) and (b) illustrates this point. One other observation made in the *p–n* heterojunction was that at approximately 1 eV strong photoabsorption in the Cu_2S created electron–hole pairs there and had a marked effect on the photoresponse.

High impurity concentrations in CdS crystals have been used to speed up their response at room temperature. The effect of high impurity concentrations is to shift the temperature break point for thermal quenching of photoconductivity to lower temperatures. This permits the superlinear behaviour in suitably doped CdS crystals to occur at room temperature and thus on removal of exciting radiation release of holes from sensitizing centres occurs immediately and consequently electrons recombine at centres with much higher capture cross-sections. The electron lifetime is then observed to decrease with time of decay and the desired effect of fast response is achieved. It is seen that the response time of CdS, which in pure material is fastest at 100°C, can be adjusted to be as fast as intrinsic CdSe at room temperature by the addition of large impurity concentrations. Skarman[42] has shown experimentally that very much faster response times can be achieved with photoconducting thin films of CdS and also CdSe simply by control in the deposition to remove undesirable defects.

5.2.2. *Cadmium Selenide and CdS–CdSe Solid Solutions*

Many of the observations on cadmium sulphide have also been made on cadmium selenide and it is not proposed in this section to reiterate ideas that have already been discussed, although the experimental results obtained for CdSe will be detailed. The behaviour of CdS–CdSe solid solutions provides an excellent link in the similarities which do exist between the constituent compounds.

Cadmium selenide has an energy gap which is in the infrared and has values derived from photoconductivity measurements of 1·714 eV ($E \perp c$) and 1·733 eV ($E \parallel c$) at room temperature.[34] At 77°K no impurity peaks are observed on the long wavelength side of the absorption edge and in this respect CdSe differs from CdS where peaks associated with bound exciton complexes were observed. Ground state A and B exciton peaks were observed along with one excited state for each exciton, $n = 2$ for the A exciton and $n = \infty$ for the B exciton. No peaks corresponding to the C exciton series were detected. The general form of the A exciton peaks in CdSe which were the only well resolved peaks was that they were broader than in CdS. At 4·2°K six new peaks occur in the photoconductivity spectra of CdSe and these are associated with two series of discrete–impurity level centres which are directly ionized to give rise to photoconduction.

The dependence of both the A and B exciton peak energies in the CdS–CdSe system on composition is very similar to that of the energy gap;[34, 43] a negative deviation of the exciton energy from a linear dependence on composition occurs and reaches a maximum value of 0·17 eV at approximately 40 mol.% CdSe.[34] The binding energy of the exciton in the solid solution ranges between 0·028 and 0·015 eV, the values in CdS and CdSe respectively. A point that should be noted is that in order for excitons to contribute to the photoconductivity they must dissociate; three mechanisms for exciton dissociation in these materials have been suggested: (a) exciton interaction with phonons, (b) exciton interaction with impurity centres and (c) exciton–exciton interaction. The latter two possibilities necessitate the motion of the exciton, whereas the first mechanism can occur under stationary conditions.

Bube[43] has analysed the thermal and optical ionization energies for sensitizing centres in CdS–CdSe solid solutions. Optical quenching measurements were made at 90°K, such a temperature being necessary to observe optical effects independently of thermal quenching in the compositions containing high proportions of CdSe; section 5.1 discusses this problem in much more detail. The threshold energy for optical quenching decreases slowly from 1·28 to 1·15 eV between CdS and 27 mol.% CdSe and from 0·76 to 0·68 eV between 50 and 100 mol.% CdSe. Thus an abrupt change in the ionization energy of the sensitizing centre occurs between 27 and 50 mol.% CdSe. A second threshold energy in the quenching spectrum of the 50 mol.% CdSe sample at 1·0–1·1 eV suggests that near this composition a change from CdS-like to CdSe-like sensitizing centres occurs. Thermal quenching effects in the CdS–CdSe solid solutions indicate a similar trend to optical quenching although the thermal quenching energy barely varies with composition and thus accentuates the change in the character type of the sensitizing centre between 27 and 50 mol.% CdSe. Figure 5.12 illustrates the behaviour of the thermal and optical

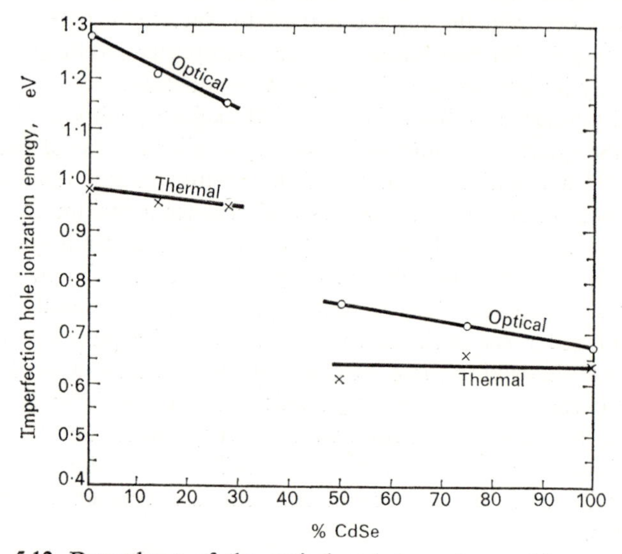

FIG. 5.12. Dependence of the optical and thermal quenching energies for sensitizing centres in undoped CdS–CdSe solutions as a function of composition.[43]

quenching energies as a function of composition in the CdS–CdSe system. The discontinuity in quenching energy is unfortunate in the application of pure CdS–CdSe solid solutions, since a continuously variable value of quench energy would have allowed easy adjustment of the response time simply by a composition change. Thus added impurities are necessary to change the response characteristics of the solid solutions and the advantages achieved over extrinsic CdS are marginal. Bube[43] concludes on comparisons with self-activated luminescence in ZnS that the sensitizing centre in the pure CdS–CdSe solid solutions is a cadmium vacancy compensated by an anion vacancy formed during the growth process. The change from CdS-like to CdSe-like behaviour occurs when the cadmium vacancy neighbours change from two sulphur and one selenium to one sulphur and two selenium plus the anion vacancy.

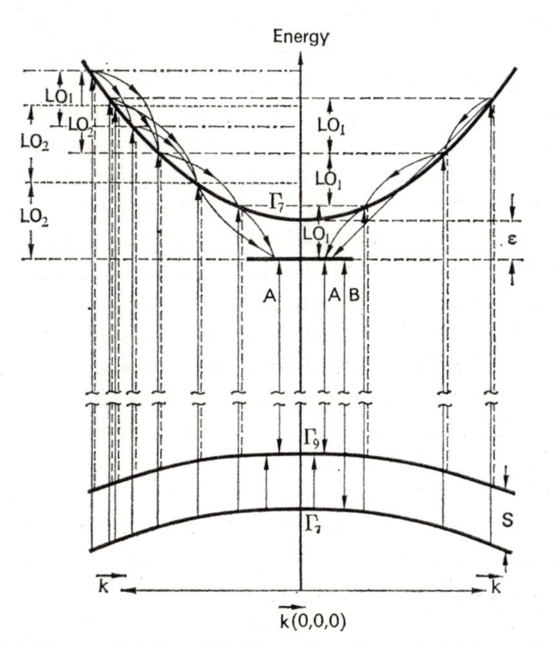

FIG. 5.13. Energy band model illustrating the upper valence bands, the conduction band and the exciton level acting as a recombination centre in a wurtzite structured material. The right-hand side illustrates the electron returning to the exciton level via LO phonon emissions of single energy in CdS. The left-hand side shows a similar return but via LO phonon emission of two energies (CdS and CdSe) in a mixed CdS_xSe_{1-x} crystal.[36]

The oscillatory photoconductivity observed in CdS is also present in CdSe and CdS–CdSe solid solutions.[37] The LO phonon energy for CdSe found from such oscillations is 0·026 eV and is derived from the slope of the A and B exciton peak energies at 4·2°K as a function of the order of the photoconductivity minimum. The ground state A and B exciton energies in CdSe are 1·826 and 1·851 eV respectively and the difference between these two energies gives a spin-orbit splitting energy of 0·025 eV, which agrees well with values determined from absorption edge measurements, see section 3.1.3. In the photoconductivity spectrum of a $CdSe_{0\cdot64}S_{0\cdot36}$ crystal at 4·2°K, minima on the high energy side of the exciton peaks can be identified with the LO phonon absorption; the phonon energies for pure CdS and CdSe have been determined from the repeat interval of the minima in these alloys. The observations in the CdS_xSe_{1-x} mixture are consistent with a model in which coupling of electrons to two different optical phonons and their additive combinations occurs and gives rise to energy levels with a high transition probability. In comparison with the pure crystals, electrons from a larger number of energy levels in the CdS–CdSe alloy will have ways of fast cascade towards an exciton level, the left-hand side of Fig. 5.13 illustrates the increased number of possibilities for the alloys compared with the single compound shown on the right-hand side. Clearly the effect of phonon coupling shortens the electron lifetime in the conduction band and exciton levels are accordingly more rapidly filled in mixed crystals than in compounds. Langer *et al.*[36] have invoked the increased population of the recombination centres (exciton levels) to explain the lower threshold energy for electron beam stimulated laser action in some CdS_xSe_{1-x} mixed crystals relative to that in either CdSe or CdS crystals as observed by Hurwitz.[44]

Bube and Ho[45] have studied the photoconductivity saturation of a $CdS_{0\cdot5}Se_{0\cdot5}$ mixed crystal at temperatures between 80 and 400°K when excited by a ruby laser or He–Ne gas laser. At low excitation intensities in a sensitive photoconductor, photogenerated holes are captured by sensitizing centres which are the dominant recombination centres for free electrons. The essential quality of the sensitizing centre over other recombination centres is their extremely low capture cross-section. However, if the excitation intensity becomes high enough, the situation will arise that all the sensitizing centres are occupied by holes and additional photoexcited holes are captured by fast recombination centres causing saturation of the photocurrent. Provided that account is taken of the density of traps which can capture electrons at saturation, it is possible to determine the density of sensitizing centres from the saturation current. The temperature dependence of the saturation photocurrent indicates the position of the Fermi-level and hence the electron trap occupancy. A density of sensitizing centres in the $CdS_{0\cdot5}Se_{0\cdot5}$ crystal of 7×10^{15} cm^{-3} was obtained and compares with values between 10^{17} and 3×10^{14} cm^{-3} in CdS measured by a similar experimental method or one using a high intensity incandescent source.[46,47] The trap density was divided into two parts, deep traps at 0·73 eV with $N_t = 3\times10^{14}$ cm^{-3} and shallow traps at 0·051 eV with $N_t = 2\times10^{16}$ cm^{-3}.

5.2.3. *Cadmium Telluride and CdTe–HgTe Solid Solutions*

De Nobel[48] in the first extensive investigations of CdTe found that the photoconductivity spectrum at room temperature peaked sharply at 1·4 eV. Subsequent investigations have tended to suggest the photoconductivity spectrum in extremely pure crystals peaks at rather higher energies. At 80 and 15°K, Vavilov *et al.*[49] observed fine structure in the photoconductivity near the absorption edge which they were able to associate with exciton transitions and LO phonon absorption. The oscillating effect in photoconductivity was correlated with the LO phonon energy of 0·022 eV. The photoconductivity spectral response is almost constant for energies higher than the forbidden band gap which is equal to 1·605 eV at 80°K. High photosensitivities in CdTe are not observed at room temperature because of the low hole ionization energy of the sensitizing centres. The energy gap in CdTe lies near the optimum for solar conversion and hence CdTe has been applied to the development of photovoltaic solar batteries; this subject is discussed in some detail in section 7.7.

The CdTe–HgTe solid solutions have shown great promise as rapidly responding photosensitive materials ever since the early observations of photoconductivity in them by Lawson *et al.*[50] Kruse[51] made an extensive analysis of single crystal slices of $Hg_{1-x}Cd_xTe$ alloys with two indium electrodes applied to a carefully prepared surface; photoconductive response in the presence of a D.C. bias was observed over the range 1 to 14 μm for different compositions at 77°K. A photo-voltage was generated in the absence of a bias in some instances and the conclusion drawn was that such voltages resulted either from carrier concentration gradients or composition gradients across the sample. Verie and Ayas[52] have successfully prepared *p–n* homojunctions by the diffusion of mercury on interstitial sites in *p*-type $Cd_xHg_{1-x}Te$ single crystals. These junctions exhibit the expected rectification properties, electroluminescence and photovoltaic effect. In the investigation, compositions for x between 0·28 and 0·15 showed photoresponses between 3 and 17·5 μm at 77°K. The structures are extremely stable in relation to the ambient atmosphere and have a fast response time of certainly less than 50 ns.

Almasi and Smith[53] used an alternative approach to observe the photovoltaic effect in $Cd_xHg_{1-x}Te$ solid solutions with a heterostructure. The heterostructure is prepared by solid state diffusion from a mixture of CdTe and HgTe powders surrounding a wafer of single crystal CdTe. Variation in the HgTe concentration permits the photosensitivity to be shifted from 0·8 to 15 μm. Microprobe analysis indicated a concentration gradient over the diffused region which can be as wide as 100 μm and the outer layer of the diffused wafer is pure HgTe. The photovoltage is dependent upon the diffusion time employed to produce the arrangement and can be as high as 500 mV.

5.2.4. *Zinc Chalcogenides*

The photoconductive and other properties of ZnO have been the subject of extensive review articles by Heiland.[54, 55] The same basic problems of slow response at low excitation intensities are observed with ZnO as with other II–VI photoconductors, although at high excitation intensities both rapid response, 0·1 ms, and high gains, 10^5, are obtained. Surface absorption creates difficulties in powder samples of charge retention which is important in image processing applications. Zinc oxide has been subjected to some rather unusual studies in relation to its photosensitivity which have perhaps been provoked by its successful application in the Electrofax process, see section 7.3. Not only have lattice defects in the bulk been used to sensitize a high photoconductivity but also both chemisorbed oxygen and organic radicals on the surface of microcrystalline ZnO powders have provided low capture cross-section recombination centres.[56] The absorbed centres permit the spectral sensitivity to be extended to regions at considerably longer wavelengths than the absorption edge. Hotchkiss[57] has observed two photon absorption from a ruby laser beam to produce simultaneously photoconductivity and fluorescence at room temperature. The photon energy of the ruby laser beam was 1·8 eV and which generated a photocurrent in the ZnO with a band gap of 3·1 eV and fluorescent emission of 2·5 eV energy. The efficiency of the process was very low and in order to observe the pulse laser effects in the photocurrent the D.C. background had to be completely filtered out.

The potentialities of the photoconductivity in zinc sulphide have not been fully realized, principally because its peak sensitivity occurs in the ultraviolet (3·69 eV) at room temperature and is outside the visible spectral region generally of interest in photoconductors. Perhaps the most interesting feature of the photosensitive behaviour observed in ZnS is the anomalously high photovoltages building up to as much as 500 V for crystals containing successive layers of cubic and hexagonal structure.[58] Neumark[59] has accounted for this behaviour with the assertion that spontaneous polarization of hexagonal ZnS causes internal fields in these crystals with opposing fields in cubic and hexagonal regions. If one then considers adjacent cubic and hexagonal sections, the opposing fields create differences in carrier concentration and a photovoltage results at the barrier between the two regions. The photovoltages from a series of these barriers are additive and hence anomalously high photovoltages are observed across the crystal. Confirmation of Neumark's theory has come from studies of the sign of the photovoltage relative to the sense of the cubic [111] axis along which transitions to the hexagonal structure occur.[60] The conclusion was that the permanent dipole in the hexagonal regions was such that the Zn (111) face was negative with respect to the S ($\bar{1}\bar{1}\bar{1}$) face. More recently photovoltages of up to several hundred volts have been observed along the surface of thin ZnS films evaporated at an angle to the substrate.[61] The photovoltage obtained is directly related to the exciting intensity, although the resistance after an initial decrease by a factor of 4 when illuminated by a tungsten lamp then remains independent of intensity over a wide range of excitation. If strongly absorbed ultraviolet radiation is added to the tungsten illumination, both the photovoltage and resistance are considerably decreased. The model proposed to explain this behaviour suggests that the ZnS layer is composed of a series of monocrystalline ZnS strips separated by ZnS material of slightly different photoresponse. Each monocrystal develops a photovoltage largely dependent upon excitation intensity and provided the interspersed ZnS is not similarly excited the photovoltages from each monocrystal add up to give a large resultant voltage. At high excitation intensities of ultraviolet radiation the interspersed

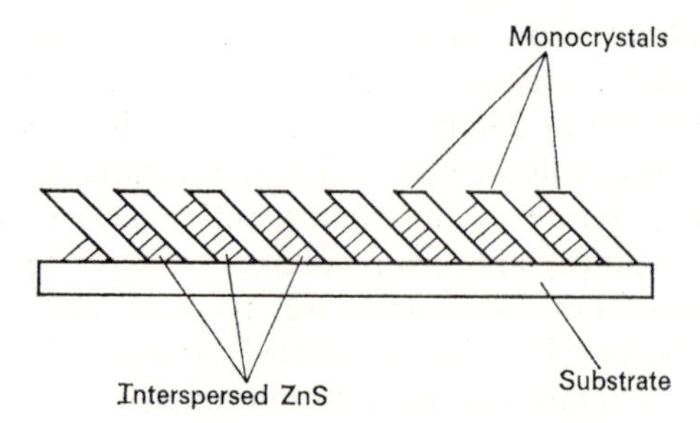

FIG. 5.14. Proposed thin film structure for a ZnS layer giving anomalous photovoltaic effect.[61]

material becomes photoconducting and causes the layer to change to a more homogeneous behaviour with a greatly reduced photovoltage and increased photocurrent. Figure 5.14 illustrates the proposed form of the evaporated layer on an appropriate substrate.

The infrared enhancement and quenching of photoconductivity in ZnS are well documented and have been used previously in section 5.1 (a) as examples of optical quenching effects.[11–13] An interesting study of the correlation between photoconductive effects and fluorescence in ZnS:Cu, Cl has helped in the identification of the nature of the emitting centre.[62] An increase in photocurrent occurs when a green emitting phosphor is excited by 365 mμm radiation whereas there is negligible photocurrent change in a blue emitting phosphor under similar excitation. The blue emitting phosphors contain a greater concentration of copper atoms and it was proposed that either the excitation occurred within a deep associated centre or the high copper density gave the electron a very short lifetime in the conduction band. Kang *et al.*[(12)] in a more detailed study of photoelectronic processes in ZnS:Cu, Cl single crystals draw rather different conclusions to those of Uchida,[(62)] in that the sensitizing centres are not thought to be the same as those responsible for luminescence. However, the fact that infrared quenching of luminescence and photoconductivity occurs at the same energy does suggest that there is some relation between the sensitizing and luminescence centres. Highly photosensitive doped ZnS films have been deposited by vapour reaction onto TiO_2 coated glass. Overall photoconductive gains of 2×10^7 have been achieved and a range of spectral sensitivities obtained by the use of different activators (Mn, As, P, Mn–Ni).[(63)]

Bube and Lind[(64)] and Bube[(65)] have made studies of the photoconductivity in cubic ZnSe crystals grown from the vapour phase and doped with halide donors and group I or group V acceptors. The general photoconductivity characteristics of ZnSe are almost identical to those discussed previously for CdS and CdSe. Appreciable infrared enhancement of photoconductivity already excited by wavelengths shorter than the absorption edge has been observed, although the wavelength of the infrared radiation must be shorter than that required for quenching. The spectral response in undoped ZnSe crystals had a peak at 462 mμm (2·68 eV) corresponding to the absorption edge. With the introduction of different acceptor and donor impurities the peak sensitivity for intrinsic effects was shifted to longer wavelengths and in heavily copper doped samples a significant additional broad peak was observed at 510 mμm (2.42 eV). The overall photoconductive gains achieved with sensitized ZnSe:Cu, Br and ZnSe:Ag, Br crystals were of the order of 10^6 but the lower carrier mobilities in ZnSe tend to give considerably lower photocurrents than in similar CdS and CdSe crystals. Decay times range from 2 to 100 ms for varying excitation intensities and a rather slow rise time of 1 s is found for low excitation intensities. Similar intrinsic photoconductive behaviour has been observed in hexagonal ZnSe crystals grown by the sublimation of a sintered ZnSe charge at 1300°C in a closed system.[66] The spectral response peaks occur at different wavelengths for polarized excitation sources and are at 437 mμm (**E**∥*c*) and 443 mμm (**E**⊥*c*) from room temperature investigations; the difference between these two peak energies gives a crystalline field interaction energy of 0·039 eV. The photosensitive peak for cubic ZnSe, as described above, occurs at an appreciably longer wavelength (463 mμm) than either of the peaks

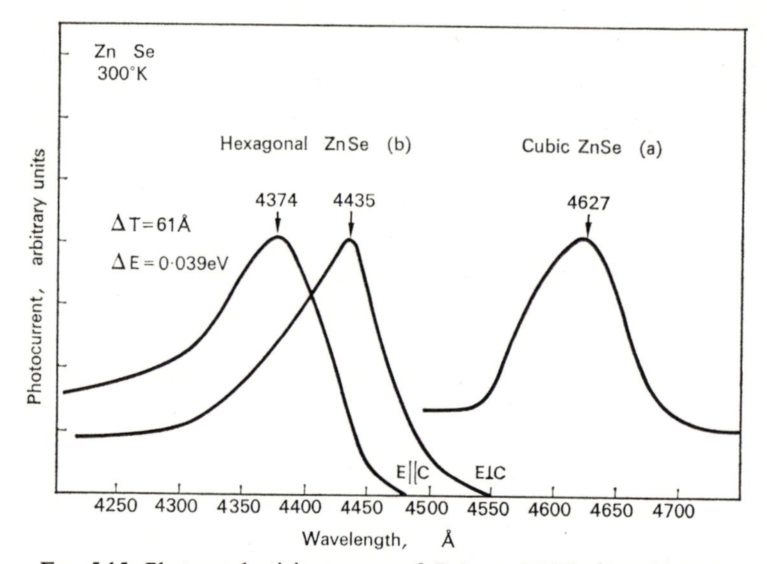

FIG. 5.15. Photoconductivity spectra of ZnSe at 300°K, (a) cubic ZnSe in unpolarized light, (b) hexagonal ZnSe in polarized light.[64]

for hexagonal ZnSe. Fig. 5.15 compares these three photoconductivity spectra.[66] Nojima and Ibuki[67] indicate they have successfully prepared *p–n* homojunctions in ZnSe and have observed the photovoltaic effect across the junction under open circuit conditions. Spectral sensitivity for this photovoltaic behaviour is a broad peak stretching between 460 and 550 mμm. Stringfellow and Bube[68] have observed the excitation of *p*-type photoconductivity in ZnSe:Cu crystals and noted the energy required is similar to that for optical quenching of *n*-type photoconductivity. They associated this form of behaviour with copper centres located 0·72 eV above the valence band, a result which is broadly in agreement with measurements on ZnS.[11, 13] Systematic studies of the photoconductive behaviour of ZnSe thin films indicate that with control of preparative conditions performance comparable to that of single crystals is achieved.[69]

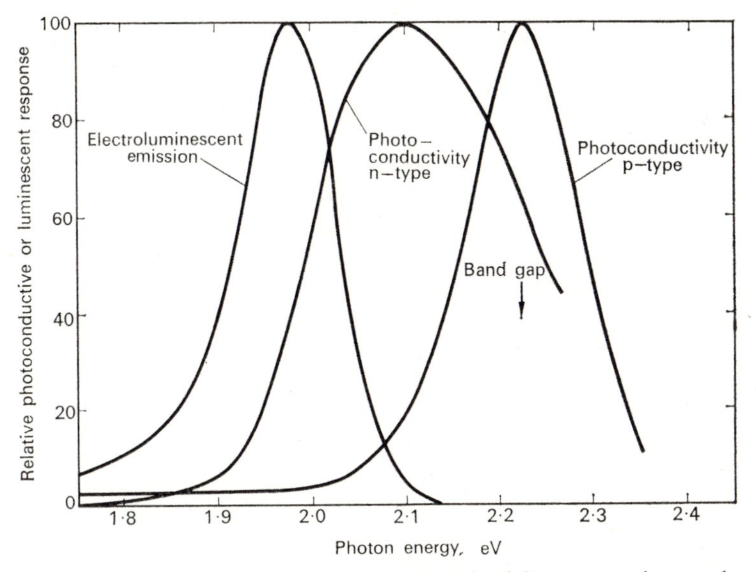

FIG. 5.16. Interrelation between the photoconductivity response in *p*- and *n*-type $ZnSe_{0\cdot 36}Te_{0\cdot 64}$ and the electroluminescent spectrum in diodes of the same composition.[71]

Photoconductive effects have been found in mixed crystals formed between ZnSe, and both ZnS and ZnTe.[70, 71] The ZnS–ZnSe:Cu system which has been investigated extensively in relation to its luminescent properties is found to have a photoconductivity spectrum very similar in form to the luminescent spectrum. The general implications are that the electrons contribute both to the conduction and the luminescent process.[70] The ZnS–ZnTe system in which light emitting

p–n junctions have been obtained exhibits very slowly decaying photoconductive properties. In an n-type $ZnSe_{0\cdot36}Te_{0\cdot64}$ crystal the photoconductivity has a sharp peak at 2·1 eV on the low energy side of the absorption edge (2·23 eV) and optical quenching effects have a threshold at 0·5 eV. Whereas a p-type copper doped crystal of the same composition has a sharp peak in the photoconductivity at 2·23 eV and an unresolved peak at 1·7 eV; the quenching spectrum has a threshold at just over 1·0 eV. The sensitizing centre in the n-type crystal was taken to be an associated self-activated centre ($V_{Zn}Al$) in which capture of a hole results in a reduction of the electron capture cross-section by 10^4 to 10^6. Aven and Garwacki,[(71)] as a result of a comparison of the photoconductivity in the p- and n-type regions and the electroluminescent emission from high resistivity $ZnSe_xTe_{1-x}$ crystals, have proposed that the probable mechanism for charge transport and light emission in such p–n junctions is based on a self-induced photoconductivity process in both the p- and n-type sides of the diodes. Figure 5.16 compares the photoconductivity response in p- and n-type $ZnSe_{0\cdot36}Te_{0\cdot64}$ and the electroluminescence spectrum emitted by a diode of the same composition.[(71)] Observations of photoconductivity in ZnTe itself have not been very extensive, although recently Takahashi *et al.*[(72)] have studied the behaviour of selected insulating crystals of ZnTe. Perhaps, one of the reasons for the lack of measurements on ZnTe is that the photosensitivity is very low at room temperature because of thermal quenching of the sensitizing centre. The photosensitivity only starts to become appreciable below 170°K rising to a maximum at 110°K. The photoconductive spectral response in ZnTe peaks at below 2·3 eV at room temperature with a shift to 2·4 eV at 77°K.

References

1. Bube, R. H., *Photoconductivity of Solids*. J. Wiley, New York, 1960.
2. Bube, R. H., *Photoelectronic Materials and Devices*, pp. 100–74 (S. Larach ed.). Van Nostrand, Princeton, 1965.
3. Bube, R. H., *Physics and Chemistry of II–VI Compounds*, pp. 655–705 (M. Aven and J. S. Prener, eds.). North Holland, Amsterdam, 1967.
4. Moss, T. S., *Photoconductivity in the Elements*, pp. 1–51. Academic Press, New York, 1952.

5. Moss, T. S., *Rept. Prog. Phys.* **28,** 15–60 (1965).
6. Nilsson, N. G., *Solid State Electronics* **7,** 455–63 (1964).
7. Ruppel, W., *Phys. Stat. Sol.* **5,** 657–82 (1964).
8. Bube., R. H., *J. Phys. Chem. Solids* **1,** 234–48 (1957).
9. Dussel, G. A., and Bube, R. H., *J. Appl. Phys.* **37,** 13–21 (1966).
10. Bube, R. H., and Henderson, D., *Physical Chemistry*, Vol. 10 (H. Eyring and W. Jost eds.). To be published by Academic Press, New York, 1969.
11. Ullman, F. G., and Dropkin, J. J., *J. Electrochem. Soc.* **108,** 154–9 (1961).
12. Kang, C. S., Phipps, P. B. P., and Bube, R. H., *Phys. Rev.* **156,** 998–1009 (1967).
13. Blount, G. H., Sanderson, A. C., and Bube, R. H., *J. Appl. Phys.* **38,** 4409–16 (1967).
14. Grabner, L., *Phys. Rev. Letters* **14,** 551–4 (1965).
15. Bube, R. H., *Phys. Rev.* **99,** 1105–16 (1955).
16. Bube, R. H., and Lind, E. L., *Phys. Rev.* **110,** 1040–9 (1958).
17. Bube, R. H., and Cardon, F., *J. Appl. Phys.* **35,** 2712–19 (1964).
18. Yoshizawa, M., *Jap. J. Appl. Phys.* **7,** 182–3 (1968).
19. Hemila, S. O., and Bube, R. H., *J. Appl. Phys.* **38,** 5258–64 (1967).
20. Nicholas, K. H., and Woods, J., *Brit. J. Appl. Phys.* **15,** 783–95 (1964).
21. Bube, R. H., *et al., J. Appl. Phys.* **37,** 21–31 (1966).
22. Dittfield, H. J., and Voigt, J., *Phys. Stat. Sol.* **3,** 1941–54 (1963).
23. Woods, J., and Nicholas, K. H., *Brit. J. Appl. Phys.* **15,** 1361–7 (1964).
24. Cowell, T. A. T., and Woods, J., *Brit. J. Appl. Phys.* **18,** 1045–51 (1967).
25. Dussel, G. A., and Bube, R. H., *Phys. Rev.* **155,** 764–79 (1967).
26. Haine, M. E., and Carley-Read, R. E., *Brit. J. Appl. Phys.* (in press, 1968).
27. Lampert, M. A., *Phys. Rev.* **103,** 1648–56 (1956).
28. Marlor, G. A., and Woods, J., *Proc. Phys. Soc.* **81,** 1013–21 (1963).
29. Tredgold, R. H., *Space Charge Conduction in Solids*, pp. 58–60. Elsevier, Amsterdam, 1966.
30. Driedonks, F., and Zijlstra, R. J. J., *Phys. Letters* **23,** 527–8 (1966).
31. Skarman, J. S., *Solid State Electronics* **8,** 17–29 (1965).
32. Crandall, R., *J. Appl. Phys.* **38,** 5425–6 (1967).
33. Handelman, E. T., and Thomas, D. G., *J. Phys. Chem. Solids* **26,** 1261–7 (1965).
34. Park, Y. S., and Reynolds, D. C., *Phys. Rev.* **132,** 2450–7 (1963).
35. Uchida, I., *J. Phys. Soc. Japan* **22,** 770–8 (1967).
36. Park, Y. S., and Langer, D. W., *Phys. Rev. Letters* **13,** 392–4 (1964).
37. Langer, D. W., Park, Y. S., and Euwena, R. N., *Phys. Rev.* **152,** 788–96 (1966).
38. Williams, R., *J. Chem. Phys.* **32,** 1505–14 (1960).
39. Grimmeis, H. G., and Memming, R., *J. Appl. Phys.* **33,** 2217–22 (1962).
40. Cusano, D. A., *Solid State Electronics* **6,** 217–32 (1963).
41. Selle, B., Ludwig, W., and Mach, R., *Phys. Stat.* **24,** K149–52 (1967).
42. Skarman, J. S., *Solid State Electronics* **8,** 17–29 (1965).
43. Bube, R. H., *J. Appl. Phys.* **35,** 586–96 (1964).
44. Hurwitz, C., *Appl. Phys. Letters* **8,** 243–5 (1966).
45. Bube, R. H., and Ho, C. T., *J. Appl. Phys.* **37,** 4132–8 (1966).
46. Bube, R. H., *J. Appl. Phys.* **31,** 1301–2 (1960).
47. Maeda, K., and Iida, S., *Appl. Phys. Letters* **9,** 92–94 (1966).
48. De Nobel, D., *Philips Res. Repts.* **14,** 361–436 (1959).

49. Vavilov, V. S., *et al.*, *J. Phys. Soc. Japan Suppl.* **21,** 156–61 (1966).
50. Lawson, W. D., *et al.*, *J. Phys. Chem. Solids* **9,** 325–31 (1959).
51. Kruse, P. W., *Applied Optics* **4,** 687–92 (1965).
52. Verie, C., and Ayas, J., *Appl. Phys. Letters* **10,** 241–3 (1967).
53. Almasi, G. S., and Smith, A. C., *J. Appl. Phys.* **39,** 233–45 (1968).
54. Heiland, G., Mollwo, E., and Stockman, F., *Solid State Physics* **8,** 191–323, Academic Press, New York, 1969.
55. Heiland, G., *J. Phys. Chem. Solids* **22,** 227–34 (1961).
56. Inoue, E., Kokado, H., and Yamaguchi, T., *J. Phys. Chem.* **69,** 767–74 (1965).
57. Hotchkiss, D. R., *J. Appl. Phys.* **35,** 2455–7 (1964).
58. Merz, W. J., *Helv. Phys. Acta* **31,** 625–35 (1958).
59. Neumark, G. F., *Phys. Rev.* **125,** 838–45 (1962).
60. Brafman, O., *et al.*, *J. Appl. Phys.* **35,** 1855–60 (1964).
61. Gagliano, A., Kramer, B., and Kallmann, H., *J. Phys. Chem. Solids* **28,** 737–40 (1967).
62. Uchida, I., *Jap. J. Appl. Phys.* **1,** 71–78 (1962).
63. Cusano, D. A., *Physics and Chemistry of II–VI Compounds*, pp. 731–6 (M. Aven and J. S. Prener eds.). North Holland, Amsterdam, 1967.
64. Bube, R. H., and Lind, E. L., *Phys. Rev.* **110,** 1040–9 (1958).
65. Bube, R. H., *Solid State Physics* **11,** 223, Academic Press, New York, 1960.
66. Park, Y. S., and Chan, F. L., *J. Appl. Phys.* **36,** 800–1 (1965).
67. Nojima, K., and Ibuki, S., *Jap. J. Appl. Phys.* **5,** 253 (1966).
68. Stringfellow, G. B., and Bube, R. H., *II–VI Semiconducting Compounds 1967 Int. Conf.*, pp. 1315–22. Benjamin, New York, 1967.
69. Dima, I., and Vasilu, G., *Phys. Stat. Sol.* **22,** K79–82 (1967).
70. Nakao, Y., *Jap. J. Appl. Phys.* **4,** 311 (1965).
71. Aven, M., and Garwacki, W., *J. Appl. Phys.* **38,** 2302–12 (1967).
72. Takahashi, R., Oshima, M., and Kobayashi, A., *Jap. J. Appl. Phys.* **5,** 339–40 (1966).

CHAPTER 6

TRANSPORT PROPERTIES

THE study of transport phenomena, by which one implies the flow of electric or thermal current resulting from an electric or thermal potential gradient, has been a rather haphazard process in the II–VI compounds. There are many reasons for this state of affairs, although it is probably most significant that the II–VI compounds found application as photoconducting and luminescent materials before any detailed understanding of their fundamental behaviour had been established. In Chapters 4 and 5 electroluminescence and photoconductivity, both of which are transport phenomena, have been presented in a way that indicates their very existence is dependent upon impurities. The impurity concentrations used in the commercially marketed devices, which apply these properties, are such as to never really require high-purity starting material. Thus, the scope for transport measurements has to some extent been limited to impure materials and in any event there is really no II–VI compound, in which there is any hope of observing intrinsic conductivity at room temperature. It is also unfortunate that this class of materials, because of the ionic character in the bond type, exhibits compensation effects for either impurity type with the result that high resistivities tend to prevail. High resistivities in group IV or III–V semiconductors suggest high purities whereas in the II–VI semiconductors such an assumption cannot generally be made.

Despite the somewhat gloomy picture painted above, there are still sufficient reliable measurements to present a systematic account of transport properties in the II–VI compounds. The chapter is divided up into an initial general section followed by four other sections on particular aspects of transport phenomena relevant to measurements made on the II–VI compounds.

6.1. General Considerations in Transport Phenomena

6.1.1. *Effective Mass*

Electronic motion in a crystalline solid is treated on the band theory of solids as an electron wave moving in a periodic potential. The allowed electron energies are obtained from the boundary conditions of the solid and the periodicity of the lattice and when speaking of particular energies the electron is treated as a wave packet. It is the effective mass tensor which relates the acceleration of the wave packet to an externally applied electric field.

$$\frac{d}{dt}\begin{pmatrix} v_x \\ v_y \\ v_z \end{pmatrix} = \begin{pmatrix} m_{xx}^{-1} & m_{xy}^{-1} & m_{xz}^{-1} \\ m_{yx}^{-1} & m_{yy}^{-1} & m_{yz}^{-1} \\ m_{zx}^{-1} & m_{zy}^{-1} & m_{zz}^{-1} \end{pmatrix} \begin{pmatrix} F_x \\ F_y \\ F_z \end{pmatrix} \tag{6.1}$$

where m_{ij} is the element of the effective mass tensor in a crystal defined by

$$m_{ij}^{-1} = \frac{4\pi^2}{h^2} \frac{\partial^2 E}{\partial K_i \partial K_j} \tag{6.2}$$

The electrons, which occupy states close to the bottom of the conduction band or the top of the valence band, determine the transport properties of a semiconductor or a semi-metal. The electron energies in the conduction band are given by a general equation which can be represented as a series of constant energy surfaces in **K**-space. In diamond-like compounds the constant energy surface is either ellipsoidal or spherical and for the former the axes of the ellipsoid in **K**-space do not necessarily coincide with the crystalline coordinate directions K_x, K_y, K_z. Thus in general an ellipsoidal constant energy surface takes the form

$$E = E_c + \frac{h^2}{8\pi^2 m_1^*}(K_1 - K_{10})^2 + \frac{h^2}{8\pi^2 m_2^*}(K_2 - K_{20})^2 + \frac{h^2}{8\pi^2 m_3^*}(K_3 - K_{30})^2 \tag{6.3}$$

where K_{10}, K_{20}, K_{30} are values of **K** at the bottom of the conduction band, E_c. The values m_1^*, m_2^*, m_3^* can only be considered constant for a limited range of **K**-values from the bottom of the conduction band. In a spherical energy surface $m_1^* = m_2^* = m_3^*$ and equation (6.3) simplifies to

$$E = E_c + \frac{h^2}{8\pi^2 m_1^*}(K - K_o)^2 \tag{6.4}$$

Spin and spin–orbit coupling effects have to be added to the above expression to describe more accurately the electron energy surfaces in the conduction band. Similar expressions are obtained for holes in

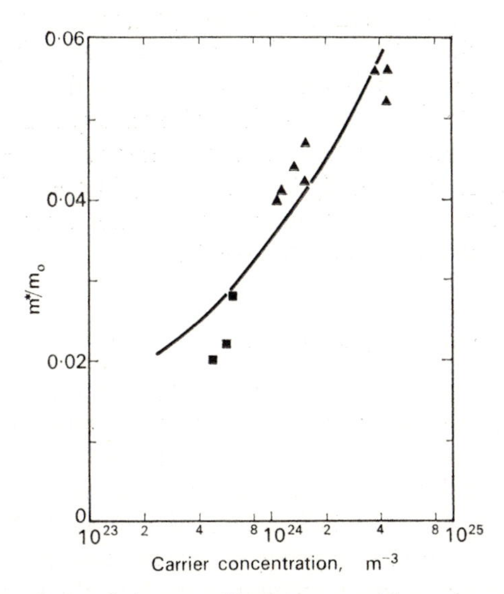

FIG. 6.1. Variation of electron effective mass with carrier concentration in HgTe.

the valence bands. Figure 6.1 illustrates the dependence of electron effective mass on carrier concentrations and thus **K** in HgTe, in which the influence of other energy bands gives a rapidly varying effective mass.

In the above discussion the effective mass tensor has been reduced to a single scalar quantity by a series of approximations based on the crystal structures pertinent to the materials in this text. The approximations cannot be strictly applied to the wurtzite structure where it is necessary to define two scalar effective masses to account for the crystalline anisotropy. Studies of CdS and CdSe have tended to indicate that the energy surfaces in the conduction band are nearly spherical, whereas the valence band effective masses are highly anisotropic.[1,2]

Devlin[3] has drawn attention to the fact that in crystals of the II–VI compounds one has to distinguish between the free carrier or high frequency effective mass m^* and the polaron or low frequency effective mass m_p. The free carrier mass is determined from free carrier absorption and Faraday rotation measurements and is the mass of the electron interacting with the lattice periodic potential. Most other experimental techniques measure the polaron mass which is the free carrier mass with an additional term due to interaction with the optical mode phonons excited in a partially ionic lattice. The interaction is determined by a coupling constant, α, generally less than unity in the II–VI compounds and as such permits the polaron and free carrier masses to be related by the approximation $m_p = m^*(1 + \alpha/6)$.

6.1.2. *Carrier Scattering Mechanisms*

In compound semiconductors there are two further important carrier scattering mechanisms in addition to the two normally observed in the elemental semiconductors such as silicon and germanium. These additional mechanisms are polar optical mode scattering and piezoelectric scattering both of which are effective in the II–VI compounds; the more common scattering processes are acoustic mode phonon and impurity scattering. A brief account of each of the scattering mechanisms will be given below.

(a) *Impurity scattering.* Ionized impurity scattering results from either deliberately introduced impurities or lattice defects generated by non-stoichiometry of the compound. The coulomb field of the ionized defect deflects the electron or hole from its paths in a way that depends on the sign of the charge on the scattering centre. Ionized impurity scattering is particularly important at low temperatures when the thermal motion of the lattice atoms is small. An expression derived for the mobility dominated by ionized impurity scattering in a non-degenerate material is

$$\mu_i = \frac{2^{7/2}(kT)^{3/2}(4\pi \varepsilon_s \varepsilon_o)^2}{\pi^{3/2}e^3 N_i(m^*)^{\frac{1}{2}}} \cdot \frac{1}{\log_e\left[1 + \frac{12\pi m^* \varepsilon_s \varepsilon_o (kT)^2}{e^2 h^2 n}\right]} \quad (6.5)$$

where N_i and n are the ionized donor and free electron concentrations respectively.[4] Observation of this form of scattering is manifested by a $T^{3/2}$ dependence of μ.

Unionized impurities and other carriers may also contribute to the scattering of carriers, although their effect will be small and only detectable at very low temperatures.

(b) *Acoustic mode scattering*. Acoustic mode scattering is conceptually difficult to represent since it is based on a wave phenomenon. The vibration of the atoms deforms the potential energy configuration of the atoms and leads to small vibrations in the energy gap. The variation in the energies of the conduction and valence band edges resulting from the vibrational motion is localized; these changes in potential energy are only effected at the expense of changes in the kinetic energy of the carriers. Hence the mobility of electrons is modified by the effects of lattice vibrations according to

$$\mu_l = \frac{2(2\pi)^{\frac{1}{2}}}{3} \cdot \frac{eh^4 \rho v_s^2}{E_{lc}^2 16\pi^4 m^{*5/2}} \cdot \frac{1}{(kT)^{3/2}} \quad (6.6)$$

where E_{lc} is the deformation potential in the conduction band, v_s is the longitudinal velocity of sound and ρ is the density. The principal feature of the acoustic mode scattering is a $T^{-3/2}$ dependence of the mobility.

(c) *Polar optical mode scattering*. Polar optical mode scattering is the most important scattering mechanism in the II–VI compounds and results from the polar interaction between charged carriers and optical phonons. It is the partially ionic character in the bond type that makes possible the induced electrostatic potential by the longitudinal optical phonons. The lattice polarization inducing the potential is a measure of the ionicity of the bond and consideration of polarization at high and low frequencies indicates the influence of ionicity on the static dielectric constant. The expression obtained for the dependence of mobility on polar longitudinal optical mode scattering is given by

$$\mu_{lo} = \frac{1}{2\alpha\omega_l}\frac{e}{m^*}\frac{8}{3\pi^{\frac{1}{2}}}\frac{F(\theta_l/T)\,[\exp(\theta_l/T)-1]}{(\theta_l/T)^{\frac{1}{2}}} \tag{6.7}$$

where $\theta_l = h\omega_l/2\pi k$, ω_l is the angular frequency of the longitudinal optical phonon, α is the polaron coupling constant related to the optical and static dielectric constants. $F(\theta_l/T)$ is a slowly varying function decreasing from a value of unity at high temperatures to 0·6 when $T = \theta_l$ and thereafter increasing steadily with θ_l/T.[5] Devlin[3] discusses the behaviour of μ_{lo} in detail and looks at the experimental techniques which satisfactorily confirm this form of carrier scattering as dominant in II–VI compounds. Low and high field Hall effect, low field transverse magnetoresistance and Nernst effect have all been used to elucidate the type of scattering.

(d) *Piezoelectric scattering.* ZnO, CdS and CdSe are II–VI crystals which are strongly piezoelectric, while the zinc blende structured members of the family also exhibit less pronounced piezoelectric properties. Piezoelectric implies electric polarization effects induced in a crystal by the application of a mechanical stress. The mechanical stress takes the periodic form of the acoustic lattice vibrations which activate polarization fields to modify the transport properties of carriers at low temperatures. Momentum conservation requirements decree that only phonons and electrons with similar wave vectors can interact and under these circumstances only at the lowest temperatures do these phonons have a significant energy to influence the transport of electrons. Hutson[6] has taken the limit of mobility due to piezoelectric scattering at 300°K to be given by

$$\mu_{lp} = 1{\cdot}44(m_o/m^*)^{3/2}(300/T)^{\frac{1}{2}}(K^2\varepsilon_o/\varepsilon_s)^{-1} \tag{6.8}$$

where K is the piezoelectric electromechanical coupling constant. The comparison between experiment and theory by Hutson for ZnO at 300°K, using $m^* = 0{\cdot}38m_o$, $\varepsilon_s/\varepsilon_o = 8{\cdot}2$ and $K^2 = 0{\cdot}074$, gave

$\mu_{Ip} = 640\ \text{cm}^2/\text{V-sec}$ and the measured value was equal to $200\ \text{cm}^2/\text{V-sec}$.

(e) *Summary*. From the above discussions the temperature dependence of mobility would seem to be a reasonable criterion for determining the dominant scattering mechanism of carriers. However, care must be taken with such interpretations since a combination of scattering mechanisms may exist and lead to a spurious result. Difficulties in impurity control in the II–VI compounds make this even more of a problem since the deep impurity levels need not be ionized until quite high temperatures and coupled with their high concentrations may result in an extremely complex mobility behaviour.

One final point which has been assumed throughout is that the phonon distribution is in equilibrium. This is justified for most scattering mechanisms at normal temperatures; however, at low temperatures the average wavelength of phonons interacting with electrons through the piezoelectric effect becomes large. Since such phonons do not interact very strongly with other phonons, their distribution can deviate appreciably from equilibrium. Deviations originating from temperature gradients inhibit the electrical properties and give rise to phonon drag effects. Hutson[(6)] has observed such effects from the Seebeck coefficient in ZnO.

6.2. Electrical Conductivity and Hall Effect

6.2.1. *Basic Concepts*

The electrical conductivity is defined as the proportionality factor between the current density and the electric field. In a semiconductor the general expression for conductivity σ is given by

$$\sigma = e(n\mu_n + p\mu_p) \tag{6.9}$$

where μ_n and μ_p are the electron and hole mobilities respectively. The carrier concentrations n and p are dependent upon the location of the Fermi energy E_F and can be represented in a simple form for intrinsic semiconductors at most temperatures and for extrinsic semiconduction at low temperatures. Expressions for these particular cases assuming parabolic energy bands are:

$$\text{(1) intrinsic } n_i = 2\left(\frac{2\pi k}{h^2}\right)^{3/2}(m_n^* m_p^*)^{3/4} T^{3/2} \exp(-E_G/2kT) \quad (6.10)$$

$$\text{(2) extrinsic } n_e = N_D^{\frac{1}{2}}\left(\frac{8\pi k}{h^2}\right)^{3/4}(m_n^* T)^{3/4} \exp(-\Delta E/2kT) \quad (6.11)$$

where N_D is the density of donor impurity states of single energy depth and ΔE is energy separation between the bottom of the conduction band and the donor level.

The above discussion assumes, in addition to parabolicity of energy bands, that the conductivity is isotropic and is open to criticism particularly in respect of the anisotropy of the wurtzite structured materials. An analysis which considers anisotropy will require a tensor to describe the conductivity and as such is beyond the scope of this text; the reader is referred to the books by Smith(7) and Ziman(8) for a treatment involving the conductivity tensor.

The Hall effect is observed as an electric field, which is created by interacting electric and magnetic fields in a conducting sample and is perpendicular to these interacting fields. The Hall coefficient R_H relates these three quantities to the free carriers contained in the conductor by

$$R_H = \frac{\varepsilon_y}{J_x B_z} = r\frac{p - nb^2}{(p + nb)^2 e} \quad (6.12)$$

where ε_y is the Hall field, J_x is the current density resulting from the applied electric field and B_z is the magnetic flux density emanating from the magnetic field; the quantities are considered in mutually perpendicular directions to observe the greatest effect. b is the electron-to-hole drift mobility ratio μ_n/μ_p and r is a factor which depends on the scattering mechanism.

In practice the conductivity of most II–VI semiconductors is dominated by one or other carrier type, hence $n \gg p$ or $p \ll n$; under these conditions the quantity $|R_H|\sigma$ is defined as the Hall mobility μ_H. The Hall mobility differs from the drift mobility μ by the scattering factor r.

The analysis of conductivity and Hall effect data is generally presented as the logarithm of conductivity or carrier concentration versus $1/T$ to determine the ionization energy for carriers and Hall mobility versus temperature on logarithmic scales to ascertain the carrier scattering mechanism.

6.2.2. *Experimental Observations*

(a) *CdS, CdSe, ZnO.* CdS, CdSe and ZnO are compounds which generally take the wurtzite crystal structure and show basically similar properties. The most striking feature which arises from conductivity and Hall effect measurements of these materials is the different regimes existing in the Hall mobility versus temperatures plots.

In CdS at temperatures between 100 and 300°K, the electron mobility follows the predicted behaviour for polar scattering, while for temperatures from 25°K down to absolute zero, piezoelectric scattering is dominant.[3, 9–11] The experimental evidence for CdS is particularly convincing because of the low density of electrons which have been obtained in "pure" crystals and Fig. 6.2 illustrates the observed data of Fujita *et al.*[9] The mobility is always seen to increase with decreasing temperature and has a value of 5×10^4 cm^2/V-sec at 1·8°K and appears to be unaffected by impurity scattering effects. In Fig. 6.2 it is seen that anisotropy of the mobility only occurs when piezoelectric scattering mechanisms are dominant. In a more recent analysis of the mobility versus temperature in CdS, Kobayashi[10] has been able to associate acoustic mode scattering with carrier transport in the temperature range 50–100°K.

The mobility behaviour with temperature in CdSe is not quite so clear cut except that polar mode scattering dominates above 100°K and there is limited evidence to support piezoelectric scattering in the purest samples at low temperatures;[3] however, in CdSe crystals containing net donor concentrations of between 10^{17} and 10^{18} cm^{-3} both ionized and neutral impurity scattering are present at temperatures as high as 200°K. If CdSe could be obtained at similar purity levels to CdS, it does seem likely that the dominance of piezoelectric scattering at low temperatures would be more clearly demonstrated. The electron mobilities observed at room temperature in the purest samples of CdS and CdSe were 350 and 650 cm^2/V-sec respectively. Narita and Shibatani[12] in investigations of Faraday rotation in CdSe observed

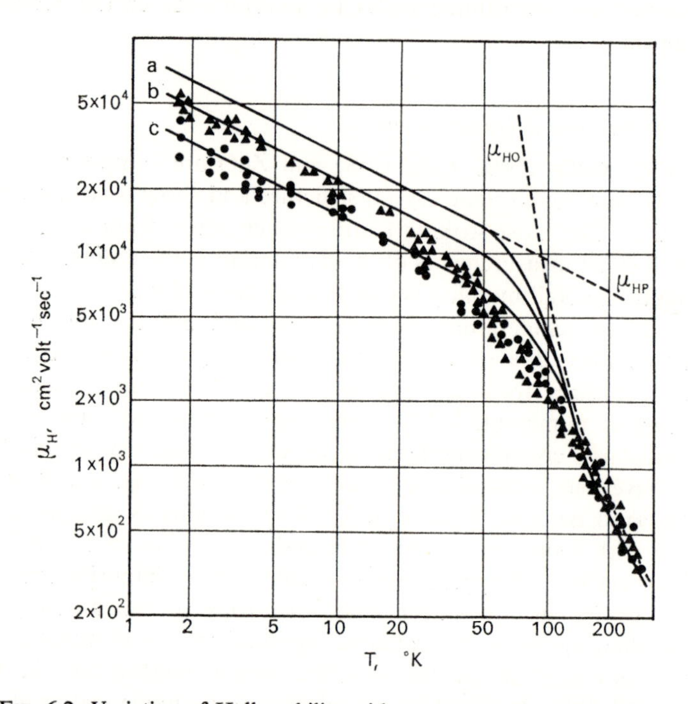

Fig. 6.2. Variation of Hall mobility with temperature in CdS. Triangles and circles indicate measurements taken with current perpendicular and parallel to the *c*-axis respectively. Curve *a* represents a hypothetical curve combining optical mode and piezoelectric data, curve *b* is the best fit of experimental data for $\mu\perp_c$ undergoing piezoelectric scattering and curve *c* is the best fit of data for $\mu\|_c$ undergoing piezoelectric scattering.[9]

that the electron mobility obtained from Hall effect measurements changed with carrier concentration N, and had values of 442 and 558 cm^2/V-sec for N equal to 0·49 and $1{\cdot}7 \times 10^{18}$ cm^{-3} respectively.

Zinc oxide has had a limited number of measurements made on single crystal material and the generally high impurity density has a strong influence on the transport of electrons at low temperatures. Watanabe *et al.*[13] observed scattering due to polar effects between 250° and 450°K, but at low temperatures with ionized impurity densities of 5×10^{18} cm^{-3} impurity scattering was taken to prevail; a room temperature electron mobility of 100 cm^2/V-sec was obtained in these measurements. In effective mass investigations of ZnO grown by

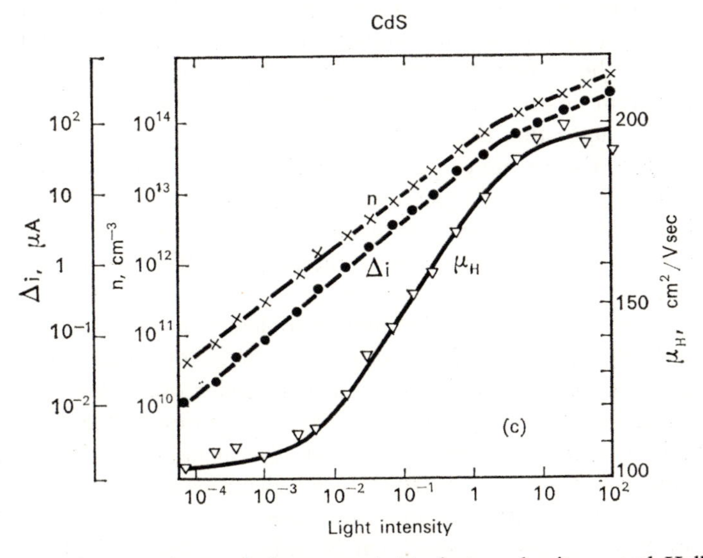

FIG. 6.3. Dependence of photocurrent Δi, electron density n, and Hall mobility μ_H, on light intensity for a CdS crystal.[15]

hydrothermal techniques, Baer[14] has found at room temperature an electron mobility of 226 cm^2/V-sec for a carrier concentration of $6{\cdot}4 \times 10^{15}$ cm^{-3}.

Observations of photo-Hall effect have been made on several semiconductors which include CdS and CdSe.[15, 16] The significant feature of the photo-Hall effect in CdS and CdSe is that the Hall mobility is dependent upon the intensity of the light excitation. In the CdS crystals at 300°K the Hall mobility increased from 100 to 200 cm^2/V-sec as the electron concentration was increased from 10^{11} to 10^{15} cm^{-3} and it has been suggested that a change in the nature of the supposed impurity scattering centres accounted for this behaviour. Figure 6.3 represents the electron concentration n, the photocurrent Δi and the Hall mobility μ_H as a function of excitation intensity.[15] Onuki and Hase[17] have also observed the electron mobility in CdS to change with wavelength of the exciting radiation and explanations for this behaviour consider the possibility of two carrier conduction at short wavelengths. Extremely large photo-Hall effect has been found in CdSe crystals and the Hall mobility increases from 100 to 900 cm^2/V-sec when the carrier concentration changes from 10^{10} to 10^{12} cm^{-3}. The apparent insensitivity of the effect to varying wavelengths has lead to the suggestion that inhomogeneous regions in the heart of the crystals give rise to built-in charged regions which are broken up by the exciting radiation.[15] The observations of Narita and Shibatani[11] mentioned above are broadly in agreement with the mobility increasing with carrier concentration, although the carrier concentrations from the different investigations are wildly different.

Spear and Mort[18] devised a method to determine both the minority and majority carrier mobilities in CdS. A pulse of 40 KeV electrons with duration of 10 ns is directed at a small area close to one electrode where free carriers are generated. Some 2 ms prior to the arrival of the excitation, a pulsed electric field of several milliseconds duration was applied across the CdS crystal; thus electrons or holes are drawn across the crystal dependent upon the direction of the field and if the time of flight across the specimen is measured, the mobility of the carriers can be determined. Room temperature hole mobilities determined by Spear and Mort ranged from 16·5 to 10 cm^2/V-sec for CdS crystals obtained from different sources. Itakura and Toyoda[19] have made measurements of the hole mobility in CdSe to which gold contacts have been applied. The gold contact is thought to produce a p-type layer at the surface of the crystal and a μ_p value of 50 cm^2/V-sec has been found for this layer.

High field effects on the electron mobility have been the subject of considerable study in CdS.[11, 20] The investigations have arisen out of attempts to clarify the interaction of electrons with acoustic waves in piezoelectric crystals. The electron mobility at room temperature is almost constant up to the threshold field, $\varepsilon_c = 1{\cdot}3 \times 10^3$ V/cm, for current saturation and thereafter it follows an inverse dependence on field up to $2{\cdot}1 \times 10^3$ V/cm; for fields greater than this the mobility decreases more rapidly than ε^{-1} and it has been suggested that at these fields impact ionization might be the cause of the change in the mobility dependence, since the Hall coefficient also starts to decrease for $\varepsilon > 2{\cdot}1 \times 10^3$ V/cm.[20, 21] Figure 6.4 illustrates the changes observed in mobility, Hall coefficient and current density as a function of field for a CdS crystal.[20]

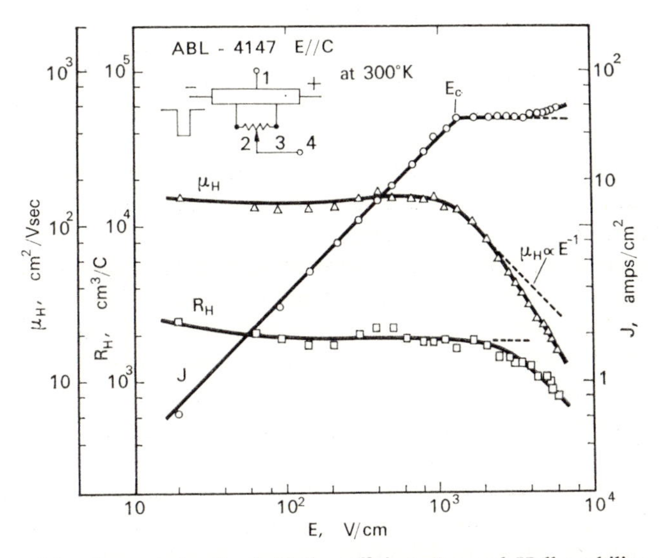

FIG. 6.4. Current density J, Hall coefficient R_H, and Hall mobility μ_H, as a function of electric field in a CdS crystal at 300°K.[20]

Kulp *et al.*[22] investigated the transport properties of electrons in CdS heavily doped with group I impurities; such crystals exhibit charge storage properties in that the dark conductivity at low temperature can be changed irreversibly by as much as 12 orders of magnitude. Very strong anisotropy in the electron mobility parallel and perpendicular to the c-axis is observed at low temperatures, $\mu\|_c$ ranging from 1 to 10 cm²/V-sec, whereas $\mu\perp_c$ is between 10 and 30 times greater than $\mu\|_c$. Resistivity as a function of temperature has an activation energy of 0·001 eV at low temperatures and coupled with a high sensitivity to changes in donor or acceptor concentrations has lead to the conclusion that the electronic conduction process occurs in an impurity band rather than in the conduction band. These conclusions are based on similar evidence to that used to substantiate impurity conduction in germanium and silicon.

The simultaneous presence of recombination centres and electron traps in these three II–VI compounds makes the interpretation of conductivity versus temperature measurements somewhat difficult, since in many instances it is not clear whether thermal equilibrium between the trap and the conduction band is achieved. Hence the energy depths of defect levels, derived from such measurements ought to be compared with those obtained from thermally stimulated current measurements, see section 5.1.4 (b). Shimizu,[23] in investigations of CdSe films, found that the substrate temperature was critical as far as both the density and the energy depth of traps introduced in the evaporated layer. Low temperature substrates (100°C) introduced a high density of shallow traps (0·14 eV) whereas high temperature substrates (200–300°C) created a high density of deeper traps (0·40 eV); substrates at 150°C provided a compromise between these two extremes. Trap depths determined from dark current versus $1/T$ and thermally stimulated current measurements agreed to within $\pm 5\%$ and had the values shown in parentheses in the previous sentence. The 0·14 eV trap has been identified with selenium vacancies and it has been suggested that possibly the 0·40 eV trap may result from a differently ionized state of the same lattice defect. Burmeister and Stevenson[24] from Hall effect measurements between 4·2 and 300°K on low resistivity *n*-CdSe have found indications of a donor level located at 0·014 eV below the conduction band and with a density of 10^{16} cm^{-3}. This shallow donor level manifested itself in the conductivity behaviour between 15 and 30°K and is clearly characteristic of the controlled preparation conditions used by these investigators. The fact that high partial pressures of cadmium were used in the preparation suggests that a cadmium interstitial might be involved with the 0·014 eV level, since the alternative possibility of a selenium vacancy has been fairly conclusively associated with the level at 0·14 eV.

Büget and Wright[25] have, from the temperature dependence of carrier concentration in several different n-type CdS crystals, found donor levels with ionization energies of 0·45, 0·63 and 0·82 eV. The same authors have also observed traps at 0·24 and 0·16 eV and the range of trapping states, that are indicated from conductivity versus $1/T$ measurements, seem to be generally in agreement with those obtained from thermally stimulated current curves as discussed by Nicholas and Woods.[26]

(b) *Zinc blende structured semiconducting II–VI compounds.* The cubic II–VI compounds which are semiconducting in nature are ZnS, ZnSe, ZnTe and CdTe. Of these compounds, ZnS has barely had its transport properties investigated, probably because its prominence as a material has resulted from its fast recombination and poorly conducting characteristics. Aven and Mead[27] have made the only really systematic investigations of conductivity and Hall mobility in low resistivity (1–10 Ω-cm) ZnS crystals produced by heavily doping with aluminium and iodine. The Hall mobility obtained was 140 cm^2/V-sec at 300°K and increased to a maximum of 300 cm^2/V-sec at 180°K. Polar optical mode scattering of electrons occurs at high temperature and ionized impurity scattering predominates below 180°K. Carrier density versus reciprocal temperature measurements indicate donor levels at 0·014 eV in iodine-doped ZnS and at 0·18 eV in Al-doped ZnS. The depth of the aluminium donor level does not compare too favourably with that obtained from thermoluminescence measurements (0·28 eV) and given in Table 4.2. However, in the thermoluminescent study it is likely that association of centres will modify the depth of the aluminium level from that in conducting ZnS. The shallowness of the iodine level is somewhat surprising in view of the appreciable degree of ionicity in the bonding in ZnS.

Devlin[3] has pointed out some observations of hole conduction in Cu-doped ZnS at 700°C with a hole Hall mobility of 5 cm^2/V-sec and a copper acceptor level located 1·2 eV above the valence band; the location of the copper acceptor level is in agreement with the energy of the infrared radiation required to quench photoconduction and luminescence in Cu-doped ZnS, see sections 4.2.2 and 5.1.3.

Zinc selenide has had its transport properties subjected to a somewhat greater level of study than has zinc sulphide. Devlin[3] reports early measurements of hole mobility in copper-doped ZnSe at 200°C which give values of 11–16 cm^2/V-sec and subsequent analysis indicates a room temperature value of 28 cm^2/V-sec; the ionization energy of the acceptor level generating holes was found to be 0·75 eV which is in agreement with the optical quenching energy for photoconduction given in Table 5.2. Aven and Segall[28] have investigated the Hall effect in ZnSe crystals which are undoped, chlorine doped, and heavily aluminium doped. The temperature dependence of the Hall coefficient indicated a chlorine substitutional level at 0·19 eV, as might be expected from thermoluminescence studies (section 4.2), a simple donor level in the undoped sample at 0·008 eV and no level at all in the aluminium-doped sample. The latter observation was suggestive of degenerate behaviour and was achieved with an aluminium concentration of $1{\cdot}5 \times 10^{20}$ cm^{-3} that appears to have overcome compensation effects. The mobility behaviour as a function of temperature is very similar to that of ZnS in that polar optical mode scattering occurs at temperatures in excess of 100°K while below this temperature ionized impurity scattering is predominant. The electron mobility in *n*-type ZnSe is between 450 and 600 cm^2/V-sec at room temperature and rises to a maximum of 7000 cm^2/V-sec in the purest crystals at 70°K.[28, 29]

The fact that normally ZnSe is *n*-type and ZnTe is *p*-type has led to several investigations of the alloys formed between these compounds because of their potentially amphoteric character. Aven and Garwacki[30] studied the transport properties of $ZnSe_xTe_{1-x}$ mixed crystals for x between 0·3 and 0·7. Crystals with $x \leqq 0{\cdot}6$ were *p*-type as grown but were converted to *n*-type by diffusing in aluminium. With x between 0·4 and 0·5 it was possible to obtain both *n*- and *p*-type modifications with resistivities of the order of 1 Ω-cm. Carrier concentration versus temperature measurements on *p*- and *n*-type alloys indicated that there were for (1) $x = 0{\cdot}6$, an acceptor level at 0·37 and a donor level at 0·01, (2) $x = 0{\cdot}5$, an acceptor level at 0·035 eV and a donor level at 0·33 eV, and (3) $x = 0{\cdot}4$, an acceptor level at 0·044 eV and a donor level at 0·33 eV. The 0·035 and 0·044 eV acceptor levels have been related to zinc vacancies and the 0·37 eV acceptor level has been associated with a noble metal impurity. The 0·33 eV donor level could well be related to aluminium on substitutional zinc sites. The electron mobilities were dominated by ionized impurity scattering at temperatures as high as 350°K probably because of the high density of ionized Al ions. The hole mobilities showed a more familiar pattern with an increase in magnitude down to 100°K and as such were indicative of polar optical mode scattering.

Zinc telluride is usually obtained in *p*-type form unless special measures are taken to remove cation vacancies. Investigations of hole mobility in ZnTe as a function of temperature have been made by several workers.[3, 28, 32, 33] Tubota[32] considered that acoustic mode scattering was predominant below room temperature since $\mu = \alpha\ T^{-3/2}$, although other investigations seemed to indicate that polar scattering governed the carrier behaviour in this temperature range.[3, 28] However, Tubota's studies were primarily concerned with impurity states in ZnTe and it is likely that the superposition of impurity effects between 100° and 300°K may have lead to a temperature dependence similar to that for acoustic mode scattering. Recent investigations by Aven[32] point conclusively to a hole mobility in *p*-type ZnTe limited by polar optical mode scattering between 80 and 500°K. Aven[32] also observed that the hole mobility at room temperature was higher in crystals with moderate carrier concentrations than those with very high and very low carrier concentrations. For crystals with high carrier concentrations (10^{19} cm^{-3}) clearly the number of charged impurity scattering centres reduces the hole mobility. However, with low carrier concentrations (10^{15} cm^{-3}) the slightly lower hole mobility is thought to be due to the diminished effect of screening by the charge carriers on the electron–phonon interaction. Figure 6.5 represents the temperature dependence of the hole mobility in ZnTe for carrier concentrations of the order of 10^{15} cm^{-3}.[32] Hole mobilities in excess of 6000 cm^2/V-sec have been observed at 40°K and it seems possible that, if detailed measurements were made below 70°K on high purity ZnTe, piezoelectric scattering may be identified. *N*-type ZnTe has been produced by Fischer *et al.*[33] using higher pressure growth technique to remove zinc vacancies at room temperature. Conductivities in the range 10^5–10^7 Ω-cm were obtained and electron mobilities ranged from

73 cm²/V-sec for unilluminated crystals to 340 cm²/V-sec for strongly illuminated crystals. These same investigators observed hole mobilities at room temperature as high as 200 cm²/V-sec in As-doped *p*-type

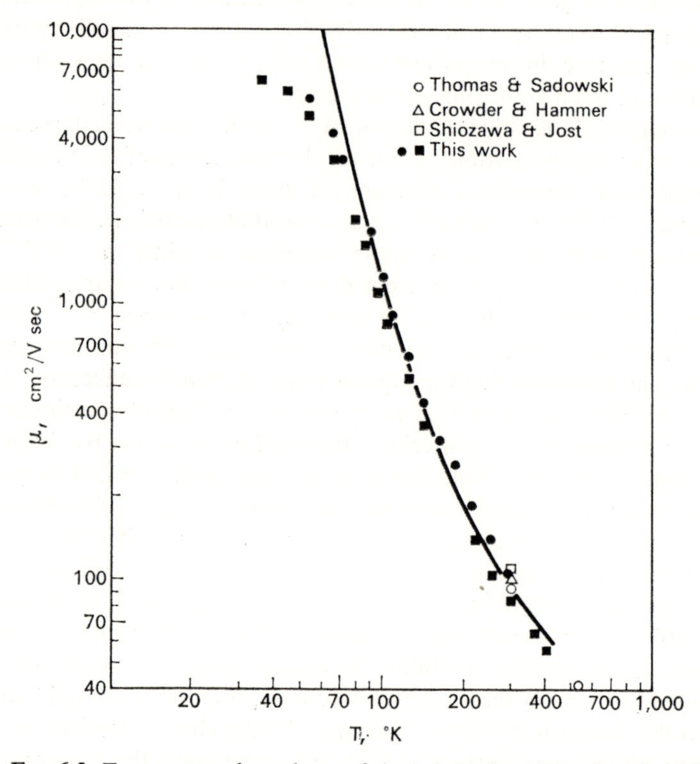

FIG. 6.5. Temperature dependence of the hole Hall mobility in ZnTe.[32]

ZnTe crystals. The indications are that the electron to hole mobility ratio in ZnTe is unusually low for a member of the II–VI family of compounds.

The analysis of Hall effect and conductivity measurements, which have been made on ZnTe, to give information about the location of defect states within the forbidden energy gap has been quite extensive.

Aven and Segall[28] obtain agreement between Hall coefficient and conductivity versus $1/T$ data for a deep copper-acceptor level with density 2×10^{16} cm^{-3} and located 0·15 eV above the valence band edge. Other deep acceptor levels result from silver and gold impurities with ionization energies of 0·11 and 0·22 eV respectively.[28] Shallow acceptor levels occur for phosphorous and lithium-doped ZnTe with an ionization energy of 0·050 eV, a value which is very close to the ionization energy of 0·057 eV of the zinc vacancy natural acceptor defect level.[32] Mazurczyk and Fan[34] have suggested that the two levels for the zinc vacancy, which is assumed to be a double acceptor, are at 0·048 eV and 0·140 eV; there would appear to be no real significance in the difference between the lower ionization energy obtained by the different investigators. Tubota[31] has tabulated activation energies of various defects in ZnTe and there is broad agreement between his values and those discussed above. Hinotani and Sugigami[35] in a study of space-charge limited current flow in semi-insulating ZnTe, determined a trap depth of 0·13–0·17 eV from the temperature dependence of the onset of the square law portion of the I–V characteristic. The depth of such a trapping level suggests that it is probably the doubly ionizable, zinc vacancy, acceptor level observed by Mazurczyk and Fan.[34]

Cadmium telluride is the only one of the cubic II–VI semiconducting compounds that can be prepared readily in *n*- and *p*-type forms. De Nobel[36] investigated the electrical properties of both *n*- and *p*-type crystals of CdTe. The effects of defect states lead to a fall in electron-mobility below 170°K suggestive of ionized impurity scattering and a maximum Hall mobility of 1200 cm^2/V-sec was observed at 170°K in *n*-type samples. The temperature dependence of electron mobility for higher temperatures ($\mu = \alpha T^{-2/5}$) did not clearly indicate one particular scattering mechanism, and de Nobel suggested a combination of acoustic mode and polar optical mode scattering to describe this behaviour. Similar conclusions were drawn by de Nobel as to the scattering of holes in *p*-type CdTe. Segall *et al.*[37], in an extensive study of *n*-type CdTe prepared under a variety of growth conditions, were able to conclude that in the purest samples polar optical mode scattering was predominant down to 100°K. Below that temperature ionized impurity effects became important and caused the mobility to eventually decrease giving a maximum value at 30°K of 57,000 cm^2/V-sec. If the carrier concentration is deliberately increased either

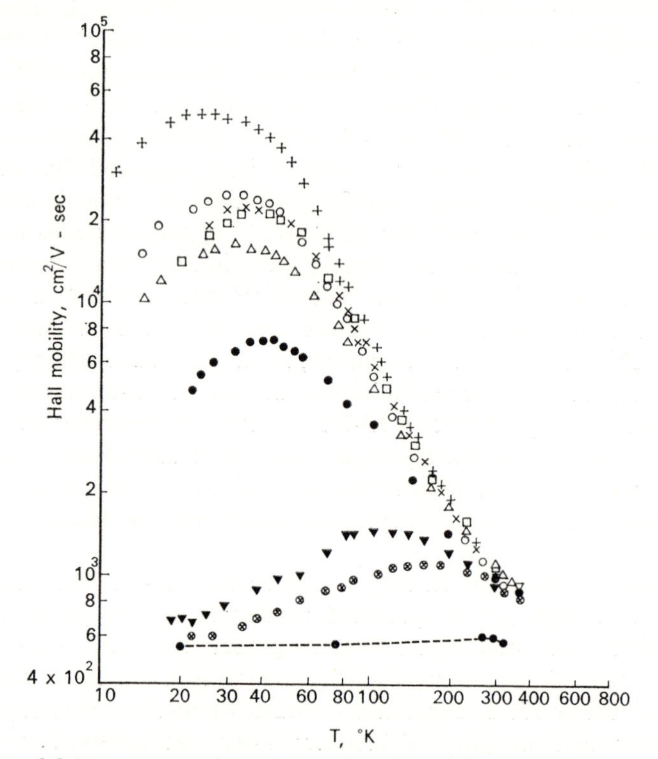

FIG. 6.6. Temperature dependence of Hall mobility in various *n*-type CdTe samples. +, ○, ×, □, △, indicate samples of multiple zone refined, undoped CdTe; ● indicates sample of multiple zone refined CdTe, annealed in excess Cd; ▼ Nd, ⊗ In, and –●– I doped samples.[37]

by annealing in excess Cd or by the introduction of donor impurities such as Nd, In or I, the mobility at low temperatures decreases by a startling amount. In fact, in iodine-doped CdTe with a carrier concentration of 2×10^{18} cm^{-3} the electron mobility barely changes between 20 and 300°K from a room temperature value of 550 cm^2/V-sec.

Figure 6.6 illustrates the mobility versus temperature behaviour obtained by Segall *et al.*[37] for a range of differently prepared *n*-type CdTe crystals. Room temperature electron and hole mobilities in CdTe up to 1050 and 80 cm^2/V-sec respectively have been observed by different investigators.

The location of defect levels within the forbidden energy gap of CdTe have been studied extensively. The interpretation of Hall coefficient versus $1/T$ data indicate that Ag, Cu and Au impurities result in acceptor levels with ionization energies between 0·30 and 0·35 eV; these ionization energy values are somewhat larger than are obtained for the same element in ZnTe.[36] De Nobel observed donor levels at 0·022 eV below the conduction band edge in both indium-doped CdTe and CdTe heat treated under high cadmium pressures; substitutional indium and cadmium interstitials were associated with these donor levels. Evidence for defect acceptor levels associated with cadmium vacancies also exists, although the location of these levels is not unambiguously determined;[36, 39] it does seem likely that this natural defect exists in both singly and doubly ionized states and is characteristic of the II–VI compounds generally. Lorenz and co-workers[38, 39] have postulated on the basis of Hall effect and conductivity measurements the existence of a double acceptor centre in CdTe heat treated with an excess of the cation component or subjected to high energy electron bombardment. The doubly ionized state of this centre is located some 0·06 eV below the conduction band maximum. The identification of a particular defect with this double acceptor level has not been made with any certainty although the fact that short heat treatments or small doses of electron irradiation create the level are suggestive of some form of native defect perhaps in association with impurities. Lorenz and Woodbury[38] have also noted that a similar defect level exists in CdS at 0·09 eV below the bottom of the conduction band. This double acceptor level when doubly ionized acts as a hole trap at low temperatures and has the effect of increasing the band gap photoexcited electron mobility three or fourfold. Figure 6.7 illustrates the changes in mobility which occur when CdTe with the double acceptor centres is subjected to photoexcitation by band gap radiation. Höschl[40] in a study of CdTe crystals at unusually high temperatures has actually observed intrinsic behaviour between 820 and 1350°K in the electrical conductivity; the energy gap obtained from these measurements of 1·58 eV is in good agreement with that found from optical absorption studies.

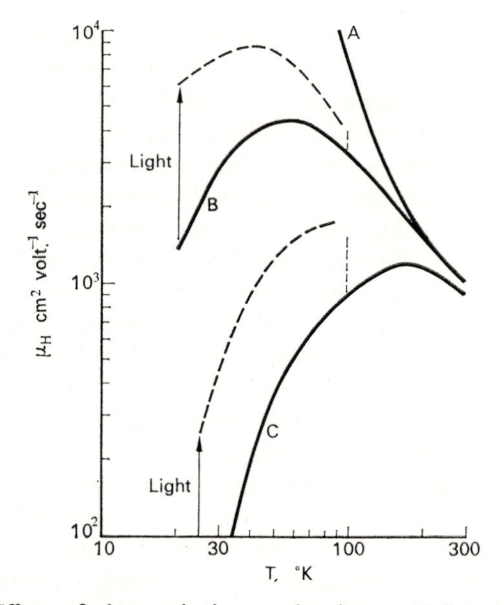

FIG. 6.7. Effects of photoexcitation on the electron Hall mobility as a function of temperature in: *A*—high purity CdTe sample. *B*—CdTe fired in Cd vapour for 30 min at 900°C. *C*—CdTe fired in Cd vapour for 285 h at 900°C.

(c) *Mercury telluride and mercury selenide.* Mercury telluride and selenide are very different from other members of the II–VI family of compounds in that they are semi-metals with something like a 0·1 eV overlap of conduction and valence band extrema. Hence the rather conventional discussion on energy states within the forbidden energy gap, related in parts (a) and (b) above, will be difficult to visualize in these two semi-metals. However, this should not detract from the interesting fundamental aspects of HgTe and HgSe that interaction between the conduction and valence bands creates small effective masses and degenerate behaviour of the electrons in the conduction band at most temperatures.

Early transport measurements on HgTe and HgSe indicated that extremely high electron mobilities were likely if the preparation conditions were controlled.(41–43) Harman(42) obtained Hall carrier mobilities at 77°K in HgTe and HgSe crystal prepared by directional freeze techniques as high as $1{\cdot}2 \times 10^5$ and 8×10^4 cm^2/V-sec respectively. Donor impurities such as Al, Ge, Si could be incorporated onto the mercury sublattice of HgSe up to concentrations of 10^{19} cm^{-3} and at these doping levels the electron mobilities at 77°K ranged from 4 to 7×10^3 cm^2/V-sec. Incorporation of the acceptor impurities, Cu, Au, and Ag, in HgSe does not have any significant effect on the carrier concentration and only produces slight scattering of electronic carriers already in existence. In HgTe almost the reverse happens in that copper incorporation produces *p*-type HgTe and a hole mobility of 90 cm^2/V-sec is observed at 77°K for a hole concentration of 3×10^{19} cm^{-3};(41, 42) the Hall coefficient generally reverses sign before room temperature is reached, principally because of the high intrinsic electron concentration and the high electron to hole mobility ratio which is of the order of 100. Other impurities, encompassing donor-, neutral- and acceptor-like, did not appreciably affect the carrier concentration although their presence significantly reduced the carrier mobility sometimes by as much as two orders of magnitude.(42)

The degenerate nature of intrinsic HgSe and HgTe, which results from the very small electron effective mass, is balanced in some respects by the stoichiometric excess of the anion generally observed. Although Blue and Kruse(44) do suggest that mercury interstitials might exist in HgSe, it would seem more likely that mercury vacancies occur and act as acceptor states. Observation of changes in the stoichiometric imbalances in HgSe and HgTe have been made by Blue and Kruse(44) and Giriat(45) respectively. Heat treatment of HgSe in a selenium atmosphere at 200°C reduces the electron concentration at 77°K to 2×10^{17} cm^{-3}.(44) Giriat(45) investigated the effects of prolonged anneals at 250–300°C in a mercury vapour atmosphere on HgTe samples and concluded that 100 hr at 250°C maximized the Hall coefficient and electrical conductivity. Both of these investigations tend to suggest that near stoichiometry is produced after suitable anneals of HgSe and HgTe.

Harman[46] has combined the Hall effect and conductivity data to analyse the dependence of Hall mobility at 300°K on carrier concentration in HgSe; the effect is quite striking in that a linear dependence is observed on a logarithmic plot and is described by $\mu = An^{-\frac{1}{3}}$ for n between 10^{17} and 10^{20} cm^{-3}. Harman *et al.*[47] have recently obtained an extremely high value of electron mobility at 4·2°K equal to 640,000 cm^2/V-sec in a single crystal sample grown at an extremely slow rate from a liquid composition slightly rich in mercury.

The scattering mechanisms for electrons in HgTe and HgSe are not so readily identified as in other II–VI compounds because of the high carrier densities. Giriat[45] observed that the mobility in HgTe satisfied a relationship of the form $\mu \propto T^{-n}$, where $n = 2{\cdot}1$ in the temperature range 330–400°K and $n = 0{\cdot}65$ near 77°K; he concluded that in the upper temperature range acoustic mode scattering predominated, whereas at temperatures near 77°K the influence of impurity scattering was appreciable. Dziuba and Zakrzewski[48] arrived at rather different conclusions for some high purity HgTe samples and considered that polar optical mode scattering was more probable at high temperatures and electron–hole scattering was dominant at low temperatures. Blue and Kruse[44] do not comment on the scattering of electrons in HgSe, although the Hall mobility versus $1/T$ curves are very similar to those obtained by Giriat[45] for HgTe; an explanation similar to that of Giriat would therefore not seem unreasonable, although it should be remembered the impurity level is not insignificant and may influence the temperature dependence of mobility over an appreciable temperature range.

It would be inappropriate to leave this section on mercury telluride and mercury selenide without some comment on the alloys formed with each other and with other chalcogenides. Harman[49] notes some early transport measurements at 4·2°K on $HgSe_xTe_{1-x}$ alloys, which were either undoped, aluminium-doped or copper-doped. In all of the HgTe specimens p-type behaviour was observed, although with aluminium doping the participation of both electrons and holes in the conduction process prevented a determination of the hole mobility. For aluminium-doped alloys the electron carrier concentration increased with x and the room temperature electron mobility fluctuated from a maximum of 9200 cm^2/V-sec at $x = 0{\cdot}25$ to a minimum of 4700 cm^2/V-sec at $x = 0{\cdot}75$. In undoped alloys the electron carrier concentration was approximately two orders of magnitude lower than in the aluminium-doped alloys and the maximum electron mobility $> 50{,}000$ cm^2/V-sec was observed for $0{\cdot}75 < x < 0{\cdot}5$. The copper-doped alloys were p-type up to $x = 0{\cdot}5$ with the hole mobility decreasing with increasing x; in the copper-doped HgSe sample an electron mobility of 55,000 cm^2/V-sec was found for $n = 3{\cdot}9 \times 10^{17}$ cm^{-3}. The $Hg_xCd_{1-x}Te$ alloys have perhaps produced the most striking transport characteristics. In this alloy system there is a composition at which the transition from semimetallic to semiconducting behaviour occurs and when this occurs a zero energy gap semiconductor exists. The first hints of the transition were seen in Lawson *et al.*'s observation of an electron mobility equal to 270,000 cm^2/V-sec at 77°K for $x = 0{\cdot}9$.[50] Subsequent investigations indicated electron mobilities at 4·2°K as high as $1{\cdot}1 \times 10^6$ cm^2/V-sec for x between 0·13 and 0·19[51,52] and it was concluded that the zero energy gap alloy had $x = 0{\cdot}17$; the electron effective mass at this composition determined from infrared measurements was $0{\cdot}006m_0$, i.e. an extremely steep curvature from the point where the conduction and valence bands touch. Other alloy systems in which the transitional behaviour from semi-metal to semiconductor have been investigated are HgTe–In_2Te_3[53] and HgSe–CdTe[54] and the same general behaviour pattern is observed to that in the HgTe–CdTe system.

6.3. Magnetoresistance and Magneto-Hall Effect

Magnetoresistance effects result from the imbalance of two opposing forces. The Lorentz force of the interacting electric and magnetic fields deflects the charge carriers and this deflection is opposed by the Hall field. The resistivity is changed because the Lorentz force is different for carriers of different velocities whereas the Hall field can only compensate the average deflection. Thus the current paths of the individual carriers have a component perpendicular to the sample axis and the resistance of the sample is modified by the applied magnetic field.

The magnetoresistance effect is defined by the ratio $\Delta\rho/\rho_o$ and for a semiconductor of single carrier type and with spherical energy surfaces is given by

$$\frac{\Delta\rho}{\rho_o} = MR_o^{\,2}\sigma_o^{\,2}B^2$$
$$= M\mu_H^{\,2}B^2 \tag{6.13}$$

where ρ_o is the resistivity in zero field, $\Delta\rho$ is the change in resistivity, R_o is the small field Hall-coefficient, σ_o is the zero field conductivity and B the applied magnetic flux density. M is a magnetoresistance coefficient which is dependent upon the scattering mechanism and if the relaxation time τ of the carriers is proportional to the energy $E(\tau \propto E^{-s})$ then

$$M+1 = \Gamma(5/2-3s)\Gamma(5/2-s)/[\Gamma(5/2-2s)]^2 \tag{6.14}$$

If τ is independent of temperature $M = 0$ and no magnetoresistance effect is observed. The above effect is the transverse magnetoresistance since it is observed transversally to the magnetic field. With large magnetic fields the resistivity takes on a saturation value ρ_∞ in which

$$\rho_\infty/\rho_o = \Gamma(5/2+s)\Gamma(5/2-s)/[\Gamma(5/2)]^2 \tag{6.15}$$

The analysis of two carrier systems is considerably more complex unless τ is constant, in which case the transverse magnetoresistance effect is represented by

$$\frac{\Delta\rho}{\rho_o} = \frac{npb\mu_p^2(1+b)^2}{(p+nb)^2}B^2 \tag{6.16}$$

The reader will appreciate, if one considers ellipsoidal energy surfaces for one or two carrier systems, the analysis is further complicated and is a subject adequately treated by Smith.[7]

The use of large magnetic fields has an influence on the Hall effect, since $\omega = eB/m^*$ is no longer small and the approximation $\omega\tau \ll 1$ is not satisfied. Under these conditions the expression for R_H in a two carriers system, in which τ is constant, is given by

$$R_H = \frac{(p-nb^2)+b^2\mu_p^2B^2(p-n)}{(p+nb)^2+b^2\mu_p^2B^2(p-n)^2}\cdot\frac{1}{e} \tag{6.17}$$

and a zero Hall coefficient occurs for

$$p = nb^2(1+\mu_p^2B^2)/(1+b^2\mu_p^2B^2) \qquad (6.18)$$

At very large fields equation (6.17) will approximate to $R_H = 1/e(p-n)$. The expression for the high field Hall effect in a system involving τ dependent on E is complex and reference on this subject should be made to Smith.[7]

Magnetoresistance measurements at 300°K in HgSe are indicative of single carrier behaviour with carrier concentrations between 3·6 and $4{\cdot}5\times10^{17}$ cm^{-3}.[46] High field Hall effect results on the HgSe sample with $3{\cdot}6\times10^{17}$ electrons cm^{-3} give an almost constant value for the Hall coefficient. Magneto-Hall effect and magnetoresistance data obtained for HgTe between 300 and 1·6° Kis rather more revealing.[46,55] Stradling and Antcliffe[55] used magnetic fields up to 80 kG and the transverse magnetoresistance $\Delta\rho/\rho_0$ at 1·6°K was observed to almost saturate at a ratio of 12 for fields in excess of 20 kG. The same investigators found that the Hall coefficient at 1·6°K reversed sign for a field of 4 kG and confirmed the simultaneous presence of high mobility electron and low mobility holes in HgTe; electron characteristics, $1{\cdot}7\times10^{15}$ cm^{-3}, $\mu_n = 168{,}000$ cm^2/V-sec; hole characteristics, $5{\cdot}4\times10^{16}$ cm^{-3}, $\mu_p = 620$ cm^2/V-sec; this gives an electron to hole mobility ratio of 300. Harman *et al.*[47] in a similar study on a large number of HgTe single crystals at 4·2°K came to somewhat different conclusions. They interpreted the results in terms of a three carrier model, two sets of electrons and one set of holes, because of the serious departures of the magnetoresistance from an inverse square law dependence on magnetic field. Figure 6.8 illustrates the behaviour of the Hall coefficient and magnetoresistance as a function magnetic field in three characteristic samples of HgTe at 4·2°K.[47] The magnitude of the Hall coefficient versus field for sample 50LB exhibits a maximum, although in fact the sign of the coefficient is negative, and is indicative of an electron majority of carriers; the maximum requires the three carrier model to explain its existence and is consistent with the band model proposed with an off-axis maximum as represented in Fig. 3.5. One other particularly interesting feature about the sample 50LB is that a hole mobility of $7{\cdot}1\times10^4$ cm^2/V-sec is observed for a hole

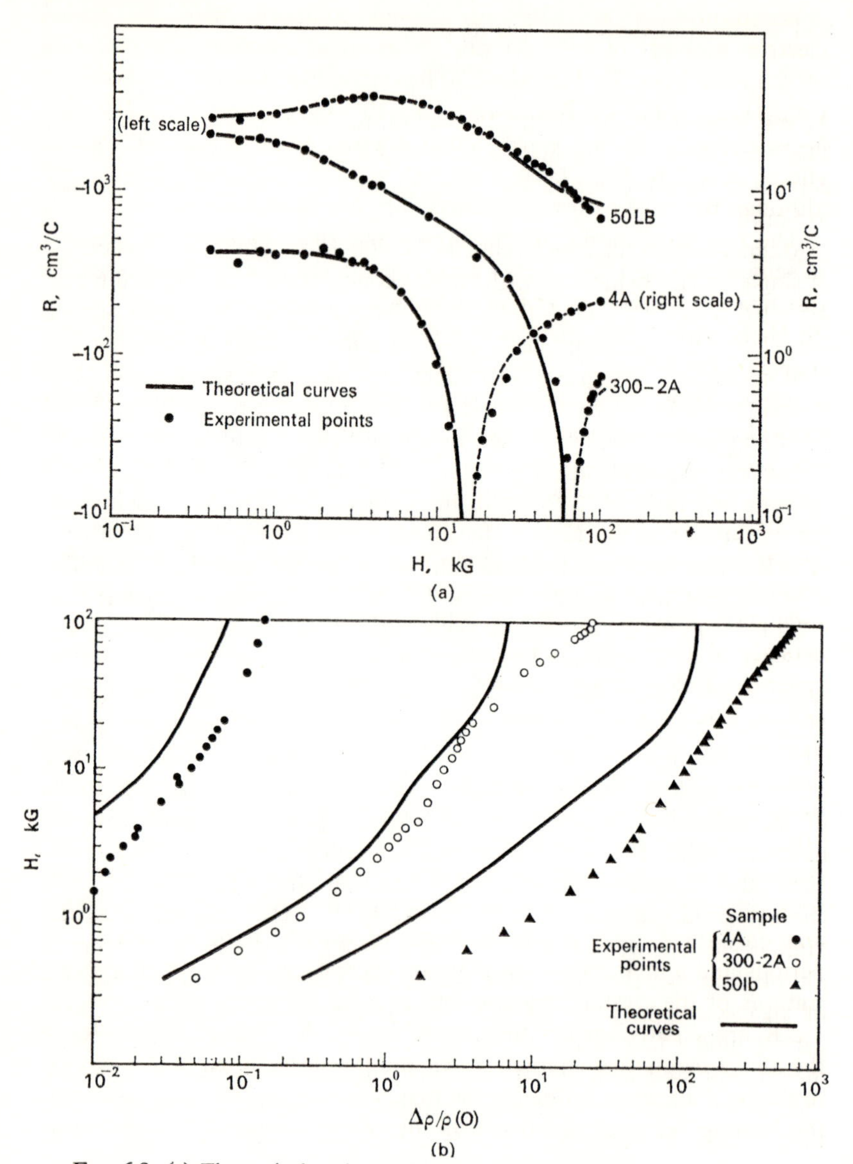

FIG. 6.8. (a) Theoretical and experimental Hall coefficient as a function of magnetic field. (b) Theoretical and experimental magnetoresistance as a function of magnetic field in HgTe at 4·2°K.[47]

concentration of $6 \cdot 2 \times 10^{14}$ cm^{-3} and compares with the higher electron mobility of $6 \cdot 5 \times 10^5$ cm^2/V-sec at an electron concentration of $2 \cdot 1 \times 10^{15}$ cm^{-3}. Harman[(49)] has made a similar analysis for $Cd_{0 \cdot 2}Hg_{\cdot 0 \cdot 8}$ Te at 4·2°K and $n = 4 \cdot 8 \times 10^{15}$ cm^{-3}, $\mu_n = 2 \cdot 7 \times 10^6$ cm^2/V-sec and $\mu_p = 76$ cm^2/V-sec. Small negative linear magnetoresistance effects at 4·2°K have been observed in both CdS and CdSe,[(24,56)] although the reasons for such effects are not fully understood.

The low temperature magnetoresistance measurements particularly in semi-metals and low energy gap semiconductors are accompanied by resistance oscillations which are commonly known as the Shubnikov–de Haas effect. The effect is based on the fact that application of a magnetic field modifies the quasi-continuum of energy states to form a set of cylindrical Landau energy surfaces parallel to the magnetic field. Each surface is separated by an energy $\hbar\omega$ and at low temperature $\hbar\omega > kT$; thus since ω is equal to eB/m^*, increases in B cause the Landau cylinders to expand and the number of energy states within the Fermi level to decrease. The energy cylinders passing through the Fermi level create periodic changes in the electronic properties observable as oscillations of magnetoresistance. Whitsett[(57)] studied the oscillatory magnetoresistance behaviour at 4·2°K of oriented single crystals of HgSe and found indications of slight deviations from a spherical Fermi energy surface; this is the kind of observation that can be made with the effect. The sample was rotated about its [110] axis so that sets of (100), (111) and (110) planes could be in turn aligned normal to the magnetic field direction. Hall coefficients were constant up to 25 kG and were used to derive the electron concentrations of the HgSe samples; the Hall mobility was in turn derived from the Hall coefficient and the low field resistance. Oscillatory magnetoresistance took on appreciable proportions for magnetic fields in excess of 50 kG and only then could general patterns of behaviour be observed. The periods and amplitudes of oscillations decrease as the electron concentrations of the crystals increase and in most instances the periods for 1·2°K are slightly smaller than those for 4·2°K. Figure 6·9 illustrates the above mentioned characteristics and it will be noted that a beating in the oscillations is strong for the magnetic field in the [111] direction; the beating becomes significant as the carrier concentration decreases and also occurs for B in the [100] direction but not in the [110] direction. Identification of the beating with slightly different frequencies which result from effective mass differences has been suggested. It has been concluded that the Fermi surface in HgSe at 4·2°K consists of a sphere with eight small protruberances in the [111] directions that do not go farther than 5% from the spherical surface. On the basis of this very small deviation from sphericity in the Fermi surface he determined an isotropic cyclotron effective mass, $m^* = eB/\omega$, from the amplitudes of oscillation at a given magnetic field and different temperatures.

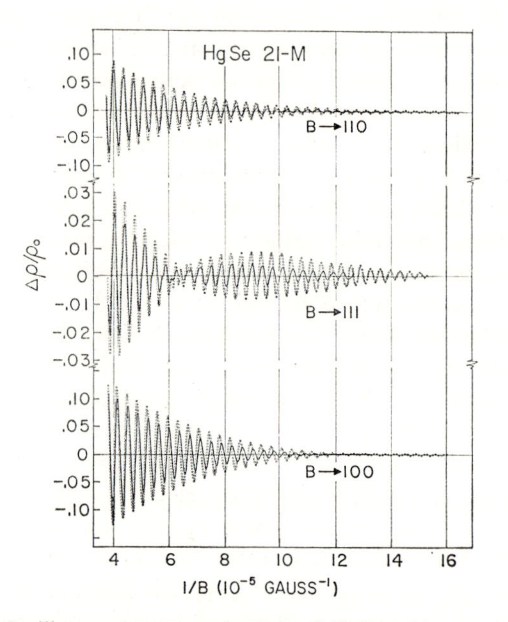

FIG. 6.9. Oscillatory magnetoresistance of HgSe with $n = 8{\cdot}46 \times 10^{17}$ electrons cm^{-3} and $\mu_H = 49{,}200\ cm^2/V\text{-}sec$ at 4·2°K. Solid curves for 4·2°K, dotted curves for 1·2°K.[57]

He found the cyclotron effective mass in HgSe to vary with carrier concentrations very much in the same manner as in HgTe, see Fig. 6.1, with m^* values of $0{\cdot}025m_0$ at $n = 10^{17}\ cm^{-3}$ and $0{\cdot}10m_0$ at $n = 10^{20}\ cm^{-3}$.

The Shubnikov–de Haas effect (oscillatory magnetoresistance) has been observed in HgTe with varying carrier concentrations at temperatures between 1·1 and 4·2°K by Stradling and Antcliffe,[55] Giriat[58] and Yamamoto.[59] The oscillations have not extended over such a large number of cycles as in HgSe because the magnetic fields available to these investigators have not been so extensive. The beating of the oscillations observed in HgSe was not present in HgTe, although oscillations in the Hall effect for HgTe are suggestive of conduction on a two-band model. Yamamoto[59] has analysed in some detail the idea of two carrier conduction and has associated two different electron mobilities with different electron concentrations at low temperatures to describe the oscillatory behaviour of HgTe samples with low carrier concentrations ($< 10^{16}$ cm^{-3}). Such a conclusion is broadly in agreement with Stradling and Antcliffe's suggestions[55] and supports Harman *et al.*'s deductions[47] based on low-temperature measurements of Hall effect and resistance as a function of magnetic field. Electron effective mass values determined from the amplitude of magnetoresistance oscillations and their dependence on temperature in HgTe range from $0{\cdot}027m_0$ for $n = 1{\cdot}2 \times 10^{16}$ cm^{-3}[59] to $0{\cdot}052m_0$ for $n = 2{\cdot}8 \times 10^{18}$ cm^{-3}.[58] Giriat also analysed the dependence of the energy of the Fermi surface in HgTe doped with large concentrations of gallium and indium and found an increase in E_F from 0·074 to 0·146 eV as n changed from 0·47 to $2{\cdot}82 \times 10^{18}$ cm^{-3}.

6.4. Other Galvanomagnetic and Thermomagnetic Effects

6.4.1. *Seebeck Effect*

In many early measurements in II–VI compounds the effective mass of carriers was determined from the Seebeck coefficient (thermoelectric power). The Seebeck coefficient S is defined for a one-dimensional temperature gradient $\partial T/\partial x$ by

$$S = \left.\frac{\varepsilon_x}{\partial T/\partial x}\right| J_x = 0 \tag{6.19}$$

where ε_x is the resultant electric field and J_x is the electrical current density along the x-axis. The expression for S in a non-degenerate n-type semiconductor is given by

$$S = -\frac{k}{e}\left\{\frac{\Gamma(s+7/2)}{\Gamma(s+5/2)} - \log_e\left[\frac{nh^3}{2(2\pi m_n^* kT)^{3/2}}\right]\right\} \qquad (6.20)$$

with the collision time $\tau = aE^s$.

Determinations of the electron effective mass in n-type CdS from the Seebeck coefficient have assumed polar optical mode scattering to be dominant above 100°K and the values obtained were between 0·2 and $0{\cdot}25m_o$;[60–62] such values are somewhat higher than those obtained from cyclotron resonance measurements with an average value of $0{\cdot}17m_o$. Morikawa[61] and Onodera[62] observed a significant contribution to the thermoelectric power at low temperatures from phonon drag. The phonon drag effects are a result of the interaction of acoustic phonons through the piezoelectric interaction and a $T^{-2\cdot5}$ temperature dependence for the phonon drag effect was observed in CdS. At 77°K the phonon drag contribution to the Seebeck coefficient was $\sim 10^3$ μV/°K and considerable anisotropy of the contribution was found for thermoelectric voltages parallel and perpendicular to the hexagonal c-axis; the electron contribution to S was approximately $0{\cdot}6 \times 10^3$ μV/°K and varied little between 77 and 300°K.

Thermoelectric power measurements in CdTe have been made on both p- and n-type samples. De Nobel[36] made a series of calculations based on such measurements at room temperature to account for the different possible scattering mechanisms. On the assumption that polar optical mode scattering was dominant, the

$$\frac{\Gamma(s+7/2)}{\Gamma(s+5/2)}$$

term in the expression for S can vary between 1·8 and 3·0 dependent upon the longitudinal optical phonon frequency; de Nobel obtained $m_n^* = 0{\cdot}14m_o$ and $m_p^* = 0{\cdot}35m_o$ taking the Γ term equal to 2 on the basis of donor ionization energy measurements. Thermoelectric power measurements in $CdTe_xSe_{1-x}$ alloys indicate the general level of voltage to be expected as the composition is varied.[63] At room temperature

S is equal to 140 μV/°K at $x = 0{\cdot}7$ and 70 μV/°K at $x = 0{\cdot}3$ for carrier concentrations of approximately 10^{19} cm^{-3}.

Several investigators have studied the Seebeck effect in HgTe and HgSe for which equation (6.20) is not strictly valid since conditions of non-degeneracy do not hold in these semimetallic compounds.(45, 46, 48, 64) Hence the values of effective mass obtained in many instances should be regarded with caution. Dziuba and Zakrzewski(48) observed the thermoelectric power in HgTe to be constant at 135 μV/°K between 77 and 300°K and obtained at electron effective mass assuming polar scattering of $0{\cdot}02m_0$; they also suggest a hole effective mass of $0{\cdot}26m_0$. Giriat(45) found a similar value for the thermoelectric power between 215 and 400°K for an electron concentration of 2×10^{17} cm^{-3}; he calculated an electron effective mass of $0{\cdot}035m_0$ using the expression for a non-degenerate semiconductor. Szymanska(64) measured the thermoelectric power between 77 and 300°K in the presence of a magnetic field and obtained values for the electron effective mass between $0{\cdot}029$ and $0{\cdot}062m_0$ for electron concentrations between 5×10^{17} and 6×10^{18} cm^{-3}. He derived an electron effective mass of $0{\cdot}018m_0$ for the conduction band minimum and concluded the behaviour of HgTe could be described basically by the Kane theory used for InSb. Harman(46), in an investigation of galvano-thermomagnetic effects in HgSe, observed the Seebeck coefficient in HgSe to decrease from 100 to 20 μV/°K as the electron concentration increases from 4×10^{17} to 3×10^{19} cm^{-3}. Although Harman makes no estimate of effective mass in HgSe, comparison with HgTe suggests the same general pattern of effective mass variation with carrier concentration and supplements Whitsetts' conclusions in section 6.3 on magnetoresistance measurements.

6.4.2. *Thermomagnetic Effects*

Only very limited measurements have been made of the thermomagnetic properties in the II–VI compounds. The potential applications of low energy gap semiconductors or semi-metals as thermomagnetic cooling materials, see section 7.8.2, has perhaps prompted some investigations of HgSe and HgTe, although the unfavourable electron to hole mobility ratio is an ultimate deterrent to their efficient usage.(46, 65, 66)

Whitsett[65] has studied the Righi–Leduc effect which is the thermal analogue of the Hall effect in HgSe crystals. The effect is particularly large in HgSe which has high electron mobility and low lattice thermal conductivity at 300°K. Thus the Righi–Leduc coefficient,

$$R_L = \frac{1}{B}\frac{\partial T/\partial y}{\partial T/\partial x}$$

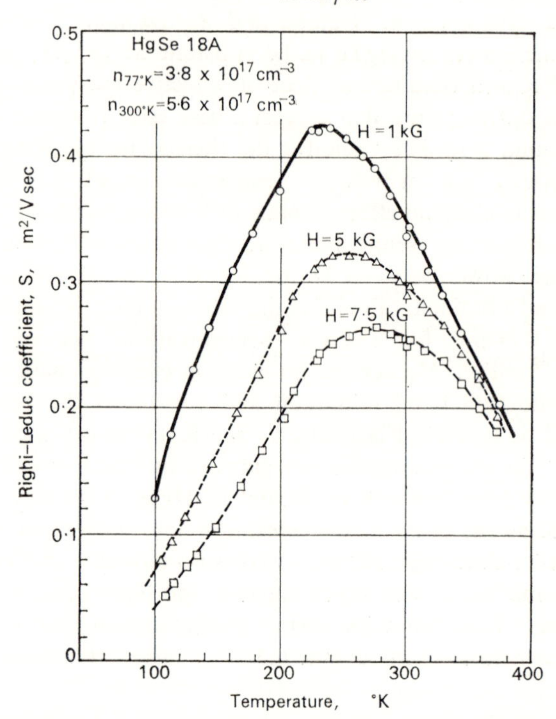

FIG. 6.10. Righi–Leduc coefficient as a function of temperature for HgSe in different magnetic fields.[65]

in a single carrier system approximates to $(\kappa_e/\kappa)\mu$, where κ_e is the electronic and κ the total thermal conductivity. The room temperature R_L values ranged between 0·27 and 0·34 m^2/V-sec as the electron concentration decreased from $5{\cdot}5 \times 10^{18}$ to $5{\cdot}6 \times 10^{17}$ cm^{-3}. Figure 6.10 illustrates the Righi–Leduc coefficient as a function of temperature for HgSe in three different magnetic fields. The coefficient reaches a maximum at just below room temperature when the electronic component of the thermal conductivity starts to become proportionately less of the total thermal conductivity.

The analysis of the Nernst effect, which is the transverse electric field generated by the presence of a thermal current in a magnetic field, in semi-metals has been undertaken by Harman.[46] The experimental results for HgSe were fitted to the theoretical expression by the choice of certain variable parameters. A maximum of the product of the Nernst coefficient N and the magnetic flux density B was observed for $\mu B = 10^8$ cm^2-G/V-sec for HgSe samples with carrier concentrations from 0·55 to $1{\cdot}4 \times 10^8$ cm^{-3}. The Nernst coefficient is defined by

$$N = -\frac{1}{B}\frac{\varepsilon_y}{\partial T/\partial x}.$$

Onodera[62] has studied the Nernst coefficient in conjunction with the Seebeck coefficient in n-type CdS to investigate the phonon drag behaviour below room temperature. He observed that screening effects by the conduction electrons give a positive phonon drag contribution to the Nernst coefficient.

6.4.3. *Helicon Oscillations and Cyclotron Resonance*

The propagation of electromagnetic waves of microwave frequencies through predominantly single carrier type semiconductors or semi-metals in the presence of a transverse magnetic field is termed helicon wave propagation. The helicon waves, which are circularly polarized, interact with the electron plasma existing within the solid provided the cyclotron polarization of the electrons created by the static magnetic field is in the same sense. The required frequency, f, for the propagation of helicons in the solid is determined by the Hall coefficient, i.e. the carrier concentration, and hence points to the obvious usefulness of the observation of helicon oscillations in a semiconductor

$$2\pi f = B|R_H|kk_z \qquad (6.21)$$

where B is the static magnetic field

$$k = \left(\frac{\pi^2 n^2}{b^2} + \frac{\pi^2 m^2}{c^2}\right)^{\frac{1}{2}} \qquad (6.22)$$

with $k_z = \pi m/c$; n, m are integers and b, c are the lateral dimensions of the sample. The decay time τ of oscillations is related to the conductivity of the sample by

$$\tau = \frac{\mu_0 \sigma B}{k^2} \tag{6.23}$$

hence the resistivity and the carrier mobility can be determined. The obvious attractive feature of the helicon oscillations is that the carrier concentration and effective mass can be determined without electrical contacts being attached to the sample. The principal requirement of the sample is that sufficient free carriers are present to keep the exciting frequency down.

Jamet *et al.*[(67)] observed the helicon effect at room temperature in an HgTe sample placed inside a resonant cavity excited in a circularly polarized mode and with a transverse static magnetic field applied. Magnetic fields in the range 0–30 kG and a frequency of 9·2 Gc/s were used in these investigations. For an HgTe sample, 0·048 mm in thickness and 3 mm in diameter, a helicon resonance was observed at 12 kG for a $\lambda/2$ dimensional resonance. The calculated electron mobility was 2×10^4 cm^2/V-sec for a carrier concentration of $0{\cdot}45 \times 10^{18}$ cm^{-3}. Figure 6.11 illustrates the power absorption as a function of magnetic field for helicon resonance in this HgTe sample. Stradling[(68)] made similar observations with an HgTe sample situated at the centre of one mirror of a confocal resonator which was part of a microwave spectrometer operating at 139 Gc/s. The experiments were performed at 20°K since the mobility was highest at this temperature. From the oscillations observed an electron effective mass of $0{\cdot}030m_0$ was calculated and an electron concentration of 7×10^{15} cm^{-3}. Gobrecht *et al.*[(69)] studied the helicon oscillations in HgSe using a secondary pick up coil wound on the single-crystal sample. They obtained a carrier concentration of $5{\cdot}25 \times 10^{17}$ cm^{-3} at 77°K which was in excellent agreement with values determined by conventional methods.

Cyclotron resonance is a further microwave absorption technique which may be used to determine the effective mass of carriers in a semiconductor at low temperatures. The sample is excited in a microwave cavity with a magnetic flux density applied normal to the direction of wave propagation and resonance occurs at an angular frequency

$\omega = eB/m^*$. For a measurable resonance to be obtained the collision lifetime τ_c must be such that $\omega\tau_c > 1$ so that the carrier executes at least part of a cycle in the magnetic field. The high carrier densities in HgTe and HgSe produce rather short carrier lifetimes and as a result they are not generally suitable for investigation by cyclotron resonance techniques. Kanazawa and Brown[70] obtained a cyclotron effective mass of $0{\cdot}096m_0$ in CdTe, which when corrected for the polar

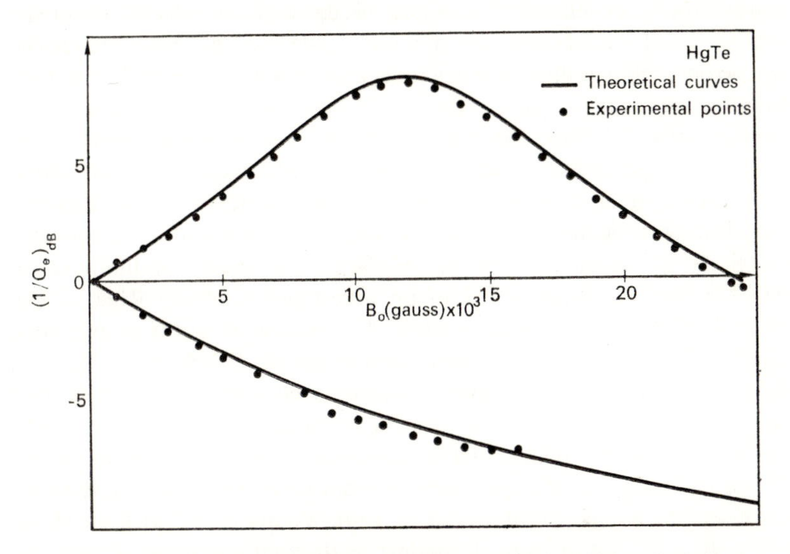

FIG. 6.11. The power absorption as a function of magnetic field for helicon resonance in an HgTe sample, 0·048 mm in thickness and 3 mm in diameter, excited at 300°K and 9·2 Gc/s.[67]

coupling gave a high frequency effective mass of $0{\cdot}090m_0$. No anisotropy in the effective mass was observed and the indications were that the conduction band consisted of a single minimum at $\mathbf{k} = 0$. Investigations of cyclotron resonance in CdS at 1·25 and 4·2°K using a 70 Gc/s signal indicate a collision time of the order of 10^{-11} sec. Considerable anisotropy in the effective mass exists and the constant energy surfaces near the conduction band minimum would appear to be oblate spheroids with transverse and longitudinal effective masses equal to

$0{\cdot}171m_o$ and $0{\cdot}153m_o$ respectively.[71] Sawamoto[72], from cyclotron resonance measurements on CdS at 1·7°K, has confirmed the above results and also obtains a hole effective mass of $0{\cdot}81m_o$.

6.5. Thermal Conductivity

The thermal conductivity in the II–VI compounds that are semiconductors is dominated by lattice effects with the electrical carriers providing only a negligible contribution. However, HgSe and HgTe, whose behaviour is characteristic of a semi-metal, have sufficient carriers to influence the magnitude of the thermal conductivity at room temperature. Carlson[41] has observed the total thermal conductivity κ_T in HgTe to vary from 0·3 to 0·027 W/cm-°K between 77 and 300°K with an electron carrier concentration at 300°K of 3×10^{17} cm^{-3}; the electron contribution to the thermal conductivity κ_e is approximately 0·008 W/cm-°K leaving a lattice thermal conductivity κ_L 0·019 W/cm-°K. Whitsett[65], in an investigation of the Righi–Leduc effect in HgSe, has obtained values of κ_T and κ_e over the temperature range 100 to 300°K. κ_T ranges from 0·099 to 0·035 W/cm-°K over the interval 100 to 300°K with an electron concentration of 6×10^{18} cm^{-3} which contributes an average κ_e of 0·02 W/cm-°K. Harman[46] has used Whitsett's results in a full analysis of galvano-thermomagnetic effects to show that the ratio of electrical to thermal conductivity for the electrons is a constant and $\kappa_L = 0{\cdot}01$ W/cm-°K at 300°K in HgSe. Aliev *et al.*[76] have studied the lattice thermal conductivity in HgSe single-crystal samples by suppressing the electronic component of thermal conductivity with a strong magnetic field. The lattice thermal conductivity in HgSe reaches a maximum at 10°K of 1·2 W/cm^{-1} °K^{-1}.

Only CdTe of the semiconducting II–VI compounds has had its thermal conductivity investigated extensively. Slack and Galginaitis[73] have analysed the effects of different impurities on the thermal conductivity in CdTe over the temperature interval 3 to 300°K. In the purest samples of CdTe for $T < 7$°K there is quite good agreement between the experimentally observed thermal conductivity and that calculated assuming the three scattering mechanisms: (1) umklapp processes, (2) crystal boundary effects, and (3) isotopes. In the temperature range

7 to 30°K there is some disagreement between theory and experiment and it has been suggested that chemical impurity or stoichiometric deficiency may give rise to an increased phonon scattering. A noticeable change in slope occurs at 30°K and above this temperature the thermal conductivity is governed by normal and umklapp scattering processes for lattice phonons. Figure 6.12 illustrates the lattice thermal conductivity in CdTe as a function of temperature for both pure and doped

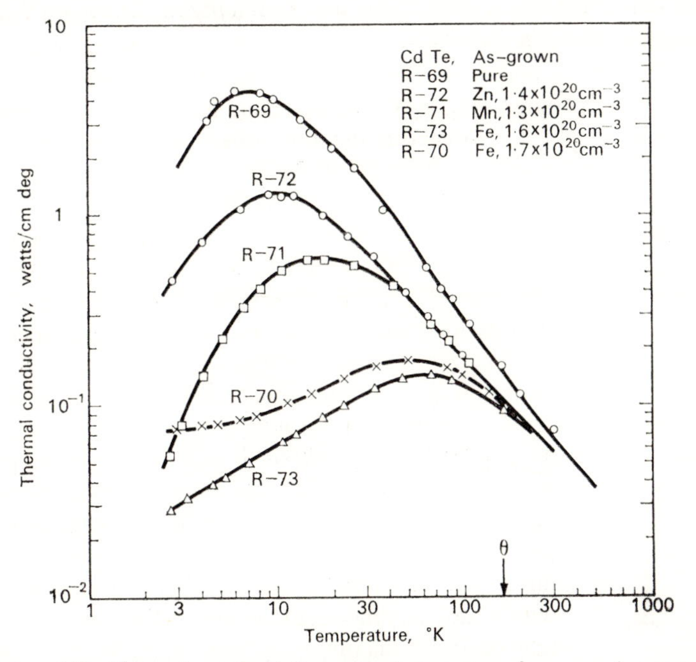

FIG. 6.12. Thermal conductivity versus temperature for pure, As-grown CdTe, and four doped samples.[73]

samples. The presence of impurities such as zinc and manganese in CdTe increased the effect of impurity scattering particularly at low temperatures with the resultant decreases in thermal conductivity. Introduction of about 1% of iron into CdTe has a quite substantial effect on the thermal conductivity below the Debye temperature ($\theta = 158$°K); in fact, at 3°K κ_L is reduced by almost two orders of magnitude. The iron is located on substitution Cd sites and gives rise to a large magnetic scattering via the arrangement of the free-ion states of the *d*-shell electrons split by the crystalline field and spin–orbit coupling. Holland[74] has observed the thermal conductivity in CdS over the same temperature range and the general features of the curve are similar to that in CdTe.

Devlin[3] reports other measurements of thermal conductivity in the II–VI compounds and these are summarized along with those already discussed in Table 6.1.

TABLE 6.1. LATTICE THERMAL CONDUCTIVITIES OF II–VI COMPOUNDS AT 300°K

Compound	κ (W/°K-cm)
CdS (‖*c*-axis)[74]	0·20
CdSe[3]	0·043
CdTe[73]	0·075
ZnS[3]	0·026
ZnSe[3]	0·19
ZnTe[3, 75]	0·18, 0·12
HgSe[46, 75]	0·11–0·030
HgTe[41, 77]	0·019–0·022

References

1. DIMMOCK, J. C., and WHEELER, R. G., *J. Appl. Phys.* **32,** 2271–7 (1961).
2. HOPFIELD, J. J., *J. Appl. Phys.* **32,** 2277–81 (1961).
3. DEVLIN, S. S., *Physics and Chemistry of II–VI Compounds*, pp. 551–609 (M. Aven and J. S. Prener eds.). North Holland, Amsterdam, 1967.
4. BROOKS, H., *Adv. Electronics Electron Phys.* **7,** 156 (1955).
5. HOWARTH, D. J., and SONDHEIMER, E. H., *Proc. Roy. Soc.* **219,** 53–74 (1953).
6. HUTSON, A. R., *J. Appl. Phys.* **32,** 2287–92 (1961).
7. SMITH, R. A., *Semiconductors*, pp. 93–188. Cambridge U.P., 1959.
8. ZIMAN, J. M., *Principles of the Theory of Solids*, pp. 182–218, 250–82. Cambridge U.P., 1964.
9. FUJITE, M., *et al.*, *J. Phys. Soc. Japan* **20,** 109–22 (1965).

10. KOBAYASHI, K., *Int. Conf. II–VI Semiconducting Compounds*, pp. 755–85. Benjamin, New York, 1967.
11. ONUKI, M., and SHIGA, K., *J. Phys. Soc. Japan* **21**, suppl. 427–30 (1966).
12. NARITA, S., and SHIBATANI, A., *J. Phys. Soc. Japan* **21**, 1218 (1966).
13. WATANABE, H., WADA, M., and TAKAHASI, T., *Jap. J. Appl. Phys.* **10**, 617–25 (1964).
14. BAER, W. S., *Phys. Rev.* **154**, 785–9 (1967).
15. BUBE, R. H., MACDONALD, H. E., and BLANC, J., *J. Phys. Chem. Solids* **22**, 173–80 (1961).
16. FOWLER, A. B., *J. Phys. Chem. Solids* **22**, 181–8 (1961).
17. ONUKI, M., and HASE, H., *J. Phys. Soc. Japan* **20**, 171 (1965).
18. SPEAR, W. E., and MORT, J., *Proc. Phys. Soc.* **81**, 130–40 (1963).
19. ITAKURA, M., and TOYODA, H., *Jap. J. Appl. Phys.* **4**, 560–6 (1965).
20. MIYAKE, T., and ONUKI, M., *Appl. Phys. Letters* **10**, 128–9 (1967).
21. YAMAMOTO, K., YANO, S., and ABE, K., *Jap. J. Appl. Phys.* **6**, 1222–5 (1967).
22. KULP, B. A., GALE, K. A., and SCHULZE, R. G., *Phys. Rev.* **140**, 252–6 (1965).
23. SHIMIZU, K., *J. Appl. Phys.* **4**, 627–31 (1965).
24. BURMEISTER, R. A., and STEVENSON, D. A., *Phys. Stat. Sol.* **24**, 683–94 (1967).
25. BÜGET, U., and WRIGHT, G. T., *Brit. J. Appl. Phys.* **16**, 1457–60 (1965).
26. NICHOLAS, K. H., and WOODS, J., *Brit. J. Appl. Phys.* **15**, 783–95 (1964).
27. AVEN, M., and MEAD, C. A., *Appl. Phys. Letters* (7, 8–10) (1965).
28. AVEN, M., and SEGALL, B., *Phys. Rev.* **130**, 87–91 (1963).
29. FUKUDA, Y., and FUKAI, M., *J. Phys. Soc. Japan* **23**, 902 (1967).
30. AVEN, M., and GARWACKI, W., *Appl. Phys. Letters* **5**, 160–2 (1964).
31. TUBOTA, H., *Jap. J. Appl. Phys.* **2**, 259–65 (1963).
32. AVEN, M., *J. Appl. Phys.* **38**, 4421–30 (1967).
33. FISCHER, A. G., CARIDES, J. N., and DRESNER, J., *Solid State Comm.* **2**, 157–9 (1964).
34. MAZURCZYK, V. J., and FAN, H. Y., *Phys. Letters* **26A**, 220–1 (1968).
35. HINOTANI, K., and SUGIGAMI, M., *Jap. J. Appl. Phys.* **4**, 731–6 (1965).
36. DE NOBEL, D., *Philips Res. Rept.* **14**, 361–430 (1959).
37. SEGALL, B., LORENZ, M. R., and HALSTED, R. E., *Phys. Rev.* **129**, 2471–81 (1963).
38. LORENZ, M. R., and WOODBURY, H. H., *Phys. Rev. Letters* **10**, 215–17 (1963).
39. LORENZ, M. R., SEGALL, B., and WOODBURY, H. H., *Phys. Rev.* **134A**, 751–60 (1964).
40. HÖSCHL, P., *Phys. Stat. Sol.* **13**, K101–3 (1966).
41. CARLSON, R. O., *Phys. Rev.* **111**, 476–8 (1958).
42. HARMAN, T. C., *J. Electrochem. Soc.* **106**, 205 (1959).
43. GOBRECHT, H., *et al.*, *J. Appl. Phys.* **32**, 2246–50 (1961).
44. BLUE, M. D., and KRUSE, P. W., *J. Phys. Chem. Solids* **23**, 577–86 (1962).
45. GIRIAT, W., *Brit. J. Appl. Phys.* **15**, 151–6 (1964).
46. HARMAN, T. C., *J. Phys. Chem. Solids* **25**, 931–40 (1964).
47. HARMAN, T. C., HONIG, J. M., and TRENT, P., *J. Phys. Chem. Solids* **28**, 1995–2001 (1967).
48. DZIUBA, Z., and ZAKRZEWSKI, T., *Phys. Stat. Sol.* **7**, 1019–25 (1964).
49. HARMAN, T. C., *Physics and Chemistry of II–VI Compounds*, pp. 767–816. (M. Aven and J. S. Prener eds.) North Holland, Amsterdam, 1967.
50. LAWSON, W. D., *et al.*, *J. Phys. Chem. Solids* **9**, 325–9 (1959).

51. Harman, T. C., *et al.*, *Phys. Rev. Letters* **7**, 403–5 (1961).
52. Rodot, M., Rodot, H., and Verie, C., *Proc. 7th Int. Conf. Phys. Semiconductors*, pp. 1237–43. Dunod, Paris, 1964.
53. Wright, D. A., *Brit. J. Appl. Phys.* **16**, 939–42 (1966).
54. Schneider, M., *C.R. Acad. Sci.* **263**, 985–8 (1966).
55. Stradling, R. A., and Antcliffe, G. A., *J. Phys. Soc. Japan* **21**, Suppl. 374–9 (1966).
56. Zook, J. D., and Dexter, R. N., *Phys. Rev.* **129**, 1980–9 (1963).
57. Whitsett, C. R., *Phys. Rev.* **138A**, 829–39 (1965).
58. Giriat, W., *Phys. Letters* **24A**, 515–17 (1967).
59. Yamamoto, M., *J. Phys. Soc. Japan* **24**, 73–81 (1968).
60. Kroger, F. A., and de Nobel, D., *J. Electronics* **2**, 190–202 (1955).
61. Morikawa, K., *J. Phys. Soc. Japan* **20**, 786–94 (1965).
62. Onodera, Y., *J. Phys. Soc. Japan* **21**, 1643–54 (1966).
63. Stuckes, A. D., and Farrell, G., *J. Phys. Chem. Solids* **25**, 477–82 (1964).
64. Szymanska, W., *Phys. Stat. Sol.* **23**, 69–73 (1967).
65. Whitsett, C. R., *J. Appl. Phys.* **32**, 2257–60 (1961).
66. Harman, T. C., *Int. Conf. II–VI Semiconducting Compounds*, pp. 982–1006. Benjamin, New York, 1967.
67. Jamet, J. P., *et al.*, *J. Phys. Soc. Japan* **21**, Suppl. 718–22 (1966).
68. Stradling, R. A., *Proc. Phys. Soc.* **90**, 175–80 (1967).
69. Gobrecht, H., Tausend, A., and Danckwerts, J., *Solid State Comm.* **5**, 551–3 (1967).
70. Kanazawa, K. K., and Brown, F. C., *Phys. Rev.* **135A**, 1757–60 (1964).
71. Baer, W. S., and Dexter, R. N., *Phys. Rev.* **135A**, 1388–93 (1964).
72. Sawamoto, K., *J. Phys. Soc. Japan* **19**, 318–22 (1964).
73. Slack, G. A., and Galginaitis, S., *Phys. Rev.* **133A**, 253–68 (1964).
74. Holland, M. G., *Phys. Rev.* **134A**, 471–80 (1964).
75. Kelemen, F., Cruceanu, E., and Niculescu, D., *Phys. Stat. Sol.* **11**, 865–72 (1965).
76. Aliev, S. A., Korenblit, L. L., and Shalyt, S. S., *Soviet Phys. Solid State* **8**, 565–9 (1966).
77. Kolosov, E. E., and Sharavskii, P. V., *Soviet Phys. Solid State* **7**, 1814–15 (1966).

CHAPTER 7

APPLICATIONS OF II-VI COMPOUNDS

THE exploitation of the II–VI compounds for commercial purposes has occurred over the past fifty years in many guises. Currently, with the surge in work on light emitting and laser materials, interest is centred on those compounds with energy gaps in the visible region of the spectrum. These and other developments represent sophisticated uses of the II–VI materials which will be discussed in greater detail when some of the more elementary applications have been considered.

7.1. Elementary Applications

7.1.1. *Photosensitive*

Both CdS and CdSe are materials employed in photoconducting cells to detect radiation in a spectral region which extends from the near infrared to gamma radiation. These highly photosensitive materials can be made to measure intensities of electron diffraction beams and of radiation at their peak spectral responses of 1·7 eV and 2·4 eV. CdS is able to handle power levels of several watts and CdSe, although it is limited to lower power levels because its sensitivity is affected by temperature increases resulting from joule dissipation, has a relatively rapid response at appreciable excitation levels of the order of 100 μs; section 5.1.2 looks at the reasons for differing response times. A response time of 100 μs, however, is still slow compared to a material like silicon and ultimately imposes a limit on the application of CdSe. CdS and CdSe have extremely high photosensitivities with maximum values of 50 A/lumen.

Photosensitive effects have also been observed for the infrared spectral region, 1–14 μm, in $Hg_{1-x}Cd_xTe$ alloys, see section 5.2.3. Both photovoltaic and photoconductive effects are present in these alloys. The nature of the alloys, which are grown by a vertical Bridgmann technique with a freeze rate of 3 mm/hr, is largely responsible for the observation of the two effects simultaneously. Non-uniformity in composition of the ingots grown by this technique results in the observation of the photovoltaic effect found in slices cut from the ingot.

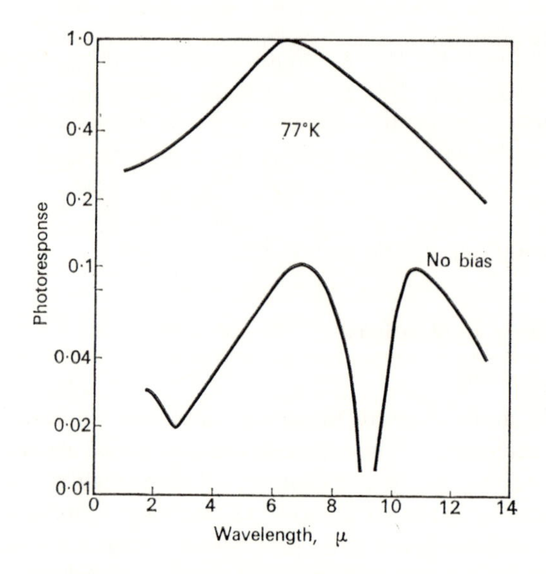

FIG. 7.1. Photovoltaic and photoconductive spectral response of an $Hg_{1-x}Cd_xTe$ sample.[1]

Figure 7.1 illustrates the photoconductive and photovoltaic effects relative spectral response for a sample with $x = 0{\cdot}2$ at 77°K. These alloys, in fact, provide useful photon detection in the 8–14 μm spectral region, which was previously inaccessible to photodetectors at 77°K. The spectral response is further extended to longer wavelengths than 14 μm, if the appropriate alloys are cooled below 77°K. Spectral response times are fast but rarely exceed 100 nsec.

7.1.2. *Luminescence*

The luminescent properties of a thin layer have an obvious potential use as a uniform source of light. This potential has been fully exploited by the lighting industry with a whole range of silicate and phosphate lamp phosphors. In these an ultraviolet discharge excites the phosphor, which in turn emits radiation in the visible region of the spectrum. The II–VI compounds although not used in the lamp industry have found application in areas such as cathode ray tube screens and warning panels in the house or in aircraft. In both these examples the luminescent material is suspended in a plastic medium and for the panels, where the activation is by an electric field, the plastic requires a dielectric constant which matches the phosphor so as to give uniform behaviour.

The spectral distribution of photoluminescent and electroluminescent emission has been extended over the whole visible region by the use of ZnS–ZnSe alloy phosphors: in these phosphors the activator material is usually copper or silver, while the coactivator is a halide. These points are more fully elaborated in sections 4.2 and 4.4. The brightness of the electroluminescent panels is optimized if the phosphor particles are kept small and uniform and the phosphor to the binder ratio is 1:2 by volume. However, the efficiency of such electroluminescence is low and rarely exceeds 1%. The lifetime of commercial electroluminescent cells varies from 1000 to 10,000 hr; it is thought that the cells decay by some form of solid state electrolysis, which results in the gradual deterioration of the cell brightness. It might appear to the reader that ZnO–ZnSe and CdS–ZnS alloys should provide suitable luminescent emitters in the visible spectrum. However, in practice the good photoconducting properties of ZnO and CdS are detrimental to their performance as luminescent materials.

7.1.3. *Optical Transmission*

Infrared optical elements have been prepared from high purity CdTe. These elements are processed from a large number of CdTe crystallites, which have been highly compressed and the resultant layer highly polished with fine abrasives. The transmission of such CdTe windows varies from 60 to 70% over the whole of the spectral region between 1 and 28 μm. Eastman Kodak Ltd. market these elements under the commercial name of Irtran 6. Irtran 6 offers the great advantage over most conventional infrared transmitters in that it is not affected by moisture and further that the same kind of optics can be used over a very wide spectral range.

7.2. Lasers

7.2.1. *General Principles*

The laser principles as a result of their historical development from masers, have been considered on a four energy level scheme. In this scheme, illustrated in Fig. 7.2, the electrons are excited from the ground

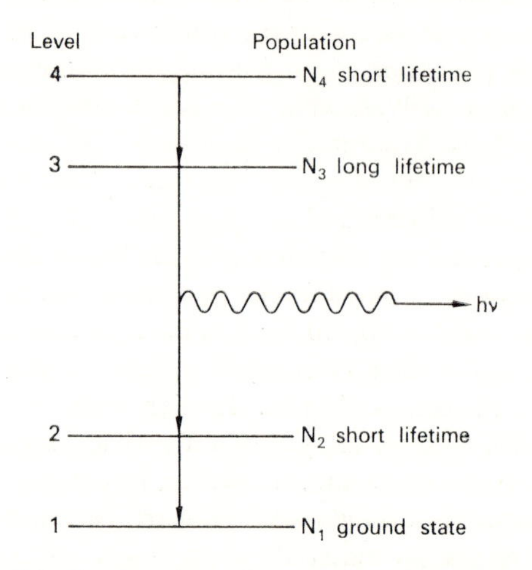

FIG. 7.2. Energy scheme for a four-level laser system.

state, level 1, to level 4, which has a short lifetime associated with it. Electrons then fall to level 3 where their lifetime is relatively long in the millisecond to microsecond range and in the process emit photons or phonons. Stimulated emission then occurs when level 2 is empty and this is usually achieved if the energy separation between level 2 and level 1 is large compared with kT thus ruling out thermal transitions. The population inversion necessary for stimulated emission is thus achieved if the pumping of electrons is sufficient to fill level 3 and make $N_3 > N_2$. The normal absorption process to take an electron in level 2 provide it with energy $h\nu$ sufficient to promote it to level 3 is now replaced by the reverse process. An incident photon stimulates an electron in level 3 to fall to level 2 and emit a photon. This stimulated emission has the same phase and direction as the initial photon. The process rapidly multiplies and continues as long as the population inversion remains. For a semiconductor level 2 corresponds to states close to the top of the valence band and level 3 to states close to the bottom of the conduction band.

In the selection of suitable laser materials a narrow spontaneous emission line is a desirable feature. A high Q-factor is also important and this is easily achieved by making the active region of the laser a part of a Fabry–Perot interferometer, which provides the directionality of the radiant beam. Nathan[2] treats this subject in very much greater detail.

7.2.2. *Pumping Mechanisms*

(a) Optical pumping is probably the most common method by which laser action generally is excited. In the method radiation, which has photon energies greater than the band gap and is frequently generated by gaseous discharges, is used to excite the semiconductor. The energy of the exciting radiation has to be matched fairly closely to the band gap or else surface absorption becomes excessive and prevents the bulk from being excited. Alternatively very small volumes have to be employed. The highest reported powers for optically pumped semiconductor lasers are 30 kW in GaAs.[3] CdSe single crystal platelets have been observed to lase when optically pumped by a $Ga(As_{1-x}P_x)$ diode laser.[4] At $x = 0{\cdot}33$ the exciting laser emits at 667 mμm when the temperature is 77°K and this wavelength falls on the high energy side of the CdSe laser emission at 692 mμm. Pulsed Xenon flash tubes have also been used to optically pump CdS_xSe_{1-x} alloys to give laser action. The duration of the stimulated emission has been only 300 ns, since sample heating causes the emission to be quenched.[5] Two photon optical absorption from solid state ion lasers has been used to pump semiconductor lasers—Nd^{3+} for GaAs and ruby for CdS.[6] The absorption constant is small and thus large volume excitation is obtained. One other group of II–VI materials in which the emission of coherent radiation has been produced by optical means is the $Hg_xCd_{1-x}Te$ alloy system.

(b) Electron beam pumping is a method frequently used for excitation of stimulated emission. High energy electrons of 20 keV or more are directed at a semiconductor sample. These electrons penetrate several microns into the material and generate electron–hole pairs as they lose energy. Approximately 10^4 electron–hole pairs are generated per incident electron and this leads to coherent emission of radiation at right angles to the incident beam. ZnS, ZnO, CdS, CdTe, CdSe and alloys of CdS_xSe_{1-x} have all been observed to lase under electron beam excitation at low temperatures (4°K and 77°K).[2,7,8] The studies on the CdS_xSe_{1-x} alloys by Hurwitz have produced stimulated emission at a range of wavelengths from the red (690 mμm) to the blue (490 mμm) with peak output powers as high as 20 W and an 11% power efficiency.

(c) *P–n* junction injection is a pumping mechanism which is unique to semiconductors. Figure 7.3 represents the form of the energy diagram of a *p–n* junction suitable to exhibit injection laser action. Both *p*- and *n*-type regions of the semiconductor need to be degenerately doped, thus locating the Fermi-level, E_F, in the valence band for the *p*-type region and in the conduction band for the *n*-type region. There exists a potential barrier to the flow of charged carriers which is removed if a voltage $V \sim E_G/e$ is applied across the junction. Under these conditions carriers can flow and recombination of electrons and holes occurs in the vicinity of the junction with the resultant emission of radiation. As the current is built up the emission increases until the threshold for laser action is obtained. An increase in light output is observed and the beam is emitted in the plane of the junction. The semiconductor sample is cut in the form of a Fabry–Perot cavity normal to the plane of the *p–n* junction. The energy of the emitted photons is generally close to that of the energy gap of the semiconductor. Unfortunately in the II–VI compounds the formation of a homogeneous *p–n* junction is a rare occurrence, since generally they are not amphoteric in character. However, *p–n* junctions have been formed in some of the II–VI alloys; ZnTe–CdTe and ZnSe–ZnTe alloys are particular examples, in which spontaneous non-coherent emission has been observed at liquid nitrogen temperatures with external quantum efficiencies

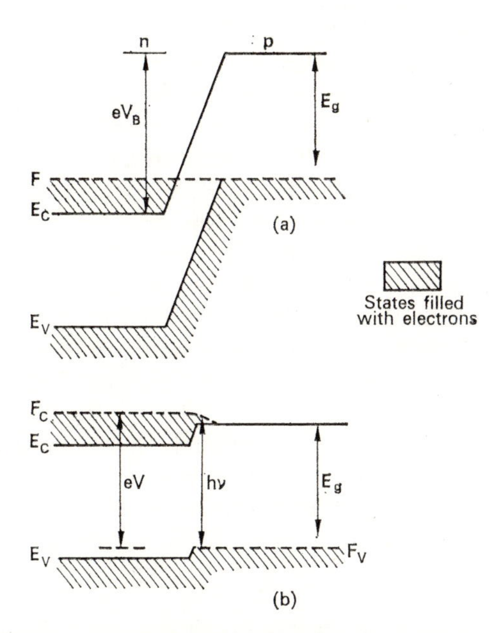

FIG. 7.3. Energy versus distance for a *p–n* junction with (a) zero applied voltage and (b) $V \sim E_G/e$.

greater than 10%.[9,10] Attempts to obtain stimulated emission in these alloys at 4°K have been unsuccessful with pulse current densities as great as 80,000 A/cm^2. It is thought that non-planarity in the junction is a possible explanation for the absence of stimulated emission. Spontaneous emission of similar efficiency has been observed by the same authors in CdTe.[11,12] High contact resistance to the *p*-type region would appear to be the reason for the absence of stimulated emission in this instance.

(d) Avalanche breakdown injection is a further pumping mechanism, with which it is possible to produce laser action. The electrons or holes in the semiconductor are accelerated by an electric field so that they have sufficient energy to generate electron–hole pairs. However, if this scheme is used in a bulk crystal the carrier velocities exceed thermal velocities and therefore prevent the formation of a degenerate distribution unless the electric field is removed, i.e. pulsed operation is necessary. Avalanche breakdown with the emission of radiation is observed in reverse biased *p–n* junctions; however, the quantum efficiency is low since electrons are injected into an *n*-type region and holes into a *p*-type region. The *p–π–p* structure has overcome this difficulty,[13] in that the holes in the *π*-region (high-resistivity *p*-type), where the field is great, initiate the avalanche breakdown. Electrons in turn are accelerated by the field in one of the *p*-regions to form a degenerate distribution and lead to laser emission in the plane of the junction. Power efficiencies in GaAs with the *p–π–p* structure are about an order of magnitude lower than in a similar *p–n* junction injection laser. So far only efficient spontaneous emission has been observed in ZnTe by avalanche breakdown at 77°K and no stimulated emission.[14] The *p–π–p* structure at least offers the opportunity to observe efficient spontaneous emission in the single carrier type members of the II–VI compounds, even if at present the poor state of the technology of contacts prevents the emission of stimulated radiation.

7.2.3. *Possible Uses*

The advantages of injection lasers such as compact size, direct conversion of D.C. electrical power to coherent radiation, ease of modulation and potentially low fabrication costs are balanced at present partially by two principal disadvantages; these are the relatively large beam angles (~ one degree of arc) and the low temperature requirement for operation at a reasonable power output. However, as efficient low power devices, *p–n* junction lasers provide an ideal solid state source of monochromatic radiation with a considerably narrower line width than that obtained from spontaneous emission, and the radiation is coherent. The line width which results from the thermal motion of the carriers known as Doppler broadening, can be further reduced by operation at low temperatures. In this respect GaAs has been exploited quite fully as an infrared source and it is to be hoped that the p–π–p structure can be similarly exploited in ZnTe, CdTe, CdSe and possibly ZnSe for the same purpose. *P–n* junctions in the ZnSe–ZnTe and ZnTe–CdTe alloys, if stimulated emission can be obtained, offer a whole range of coherent monochromatic light sources in the visible region of the spectrum. The energy of emission in semiconductor junction lasers can be varied by changes in temperature, pressure and magnetic field, hence accurate control of these factors is a great asset in laser applications. Injection lasers have been used as amplifiers and in such cases, antireflection coatings are put on the ends of the cavity to reduce the feedback and inhibit oscillation. An increase of the threshold current by a factor of 10 leads to an increase in the light intensity by a factor of 2000.

Injection lasers are capable of being both amplitude and frequency modulated. In GaAs for example, the limit to the speed of modulation lies in the detector rather than the modulator. The response time of the laser to fast rise pulses in less than 0·2 ns. There does, however, exist a time delay t_d between the onset of the current pulse and the observation of stimulated emission. The delay is caused by the time required to invert the population of carriers, which is related to the spontaneous lifetime τ and the current level I by the following expression

$$t_d = \frac{\tau ln I}{I - I_{th}} \qquad (7.1)$$

where l is the length of the Fabry–Perot cavity, n is the average carrier density and I_{th} is the current threshold for stimulated emission.[2] The delay does not inhibit the laser's response to microwaves, even if the current level falls to zero, provided the period of the waves is less than the spontaneous lifetime. Amplitude modulation has been reported at 10 Gc/s for lasers operated at liquid helium temperatures. Frequency modulation has been effected by the variation of the refractive index using ultrasonic techniques; here an upper frequency limit of 1 Gc/s is set by the requirement of uniformity in the active region of the laser. However, with the exception of liquid helium temperatures the refractive index is extremely sensitive to temperature ($\sim$ 20 Gc/s/°K), thus almost impossible temperature control is required to utilize the effect.

A rapidly responding light source like the injection laser has potential application in the measurement field for items such as carrier lifetimes in semiconductors, fluorescent decay times and detector response times. Amplitude modulation at high frequencies makes the laser attractive for communication purposes. The directionality of the laser beam compared with an incoherent light emitting diode coupled with its high overall efficiency make it an excellent prospect as an optical communications transmitter. Atmospheric attenuation limits its terrestrial application to short distances, however there are obvious possibilities in space where the attenuation is negligible and GaAs lasers have been used for this purpose in the American space programme. On the terrestrial scale a laser has been used to transmit 24 voice channels over a distance of 13 km.[15]

In the optical computer field the injection laser does not show very much promise at the present time. However, optical interaction between lasers does indicate that logical operations can be performed. Logical negation is obtained when one laser is used to quench the emission of another. Radiation from one laser is passed into the active region of the other laser at right angles to its emitting direction. The incident radiation is amplified and thus energy is absorbed from the second laser, which if it is operating near its threshold will result in quenching of the stimulated emission. This switching process has been effected efficiently by building the two *p–n* junction lasers on the same *n*-type layer and exciting them at right angles to each other.[16]

The emphasis in the discussion of possible uses of lasers has been on injection lasers. It is felt that as the technology of II–VI materials improves and approaches somewhere near the present state of affairs in GaAs, the injection, either *p–n* junction or p–π–p structure, laser with the tellurides and selenides will realize its full potential. To conclude, it seems worthwhile to mention the potential of II–VI compounds in high-power lasers. At present no four- or three-level type lasers (Nd^{3+} or ruby) have been observed in the II–VI materials. MgO is probably the most likely material since the Cr^{3+} ion gives a similar fluorescent emission to that in alumina. Rare earth ions in the II–VI compounds may also be a possibility for this type of laser.

With high power lasers, applications in the micromachining, cutting and welding fields are the most obvious. The laser range finder represents a more refined use of a high power pulsed laser.[17] A ruby laser has been used for this purpose with pulses of 40 ns duration and 4 MW peak power. The laser beam is switched with a rotating prism when it is parallel to a partially transmitting mirror. A photomultiplier is used to detect radiation after appropriate collimation and reflection by the object of interest. It is fair to say that the exploitation of the II–VI compound laser materials and their subsequent applications is at an early stage.

7.3. Electrophotography

An analogy may be drawn between electrophotography and conventional halide photography in that a latent image is formed on a sensitive layer by exposure to radiation. However, in electrophotography the sensitive layer is a charged photoconductor and hence the latent image is a charge pattern. Subsequent development uses the electrostatic attraction of the charge pattern to provide a real image with finely divided powder. Thin layers of sulphur or selenium on a metal backing provided the early electrophotographic plates. If dry development and processing is employed then the photosensitive plate can be used repeatedly.

Young and Grieg[18] discovered that photoconductive zinc oxide crystallites dispersed in a polymer resin could be used as a sensitive layer in electrophotography and in such a condition the zinc oxide is usually referred to as Electrofax. The sensitive layer is suitable for coating on paper since it is flexible and in addition has the advantages of being white, inexpensive and non-toxic. Thus with Electrofax the whole processing can be carried out on a single layer, which has advantages in respect of the definition and multicolour process requirements. It is proposed to discuss in brief in the following sections the development of Electrofax. At times such development may appear to have a slight touch of alchemy, but nevertheless the end would seem to justify the means. Amick[19] has reviewed Electrofax behaviour in much greater detail.

7.3.1. *Selection of Material for Electrofax*

Stringent materials requirements have to be applied to produce suitable Electrofax layers. The choice of zinc oxide powder is especially critical as a result of which certain empirical tests are applied to the zinc oxide powder and the Electrofax layer. Three basic tests applied to zinc oxide indicate that the highest purity oxides are most suitable for Electrofax. The tests are:

(1) Observation of the photoconductivity of a pressed pellet of zinc oxide powder indicates that the maximum surface conductivity must be greater than 10^{-8}/ohm/W/cm^2.

(2) The photoluminescence of the powder after firing in dry hydrogen at 1000°C for 5 min should be bright green under 3650 Å ultraviolet excitation.

(3) The photoluminescence of zinc oxide dispersed in silicone resin and cooled to liquid nitrogen temperatures should be lavender to peach–lavender in colour for 3650 Å excitation.

Electrofax layers are further divided into good and bad materials by a simple test in which the layer is corona charged and then the potential across it is measured. Useful layers will have saturation potentials of greater than 90 V and poor layers potentials of 40 V or less; this is a simple empirical test which has yet to be confounded. Potentials between these two values have not been observed. The type of resin used is not too critical provided that moisture is kept away from the surface of the layers. Silicone, rubber and water based compositions have all been used successfully.[19] The resin to zinc oxide ratio has to be such that the zinc oxide particles effectively touch each other without excluding the resin. Addition of selected organic dyes to Electrofax layers can increase the layer sensitivity to white light. The dyes, which are strongly absorbed on the surface of the zinc oxide crystallites, absorb a photon of visible radiation and then inject an electron into the conduction band of a crystallite. Development is effected by powders which carry a triboelectric charge, the sign of which depends on whether a positive or reversed print is required. Magnetic brush development is employed, if dry development is preferable and wet development is achieved with a dispersion of the triboelectric powder in a suitable silicone oil.

The Electrofax layers used in practice consist of a 1·25–15 μm thick coating of the photosensitive mixture on a clay coated paper backing of thickness 60–70 μm.

7.3.2. *Fundamental Features of Electrofax Layers*

It has been found despite the particle nature of the zinc oxide powder in Electrofax layers that the crystalline particles are in electrical contact. Also since the crystallite size is smaller than the depletion layer that can form in the material, there exists no real distinction between surface and volume characteristics. It is thus possible to use an energy band model for a homogeneous semiconductor to represent the case of zinc oxide in an Electrofax layer. The zinc oxide both in the dark and under illumination exhibits *n*-type conductivity, which is probably caused by a zinc excess on interstitial lattice sites. In fact, the zinc oxide crystallites accept oxygen atoms at the surface in excess of the stoichiometric value. The electrons obtained from the zinc donor ions ionize the atmospheric oxygen at the surface to produce negative oxygen ions. The effect of this process in the dark is to make the crystallite highly insulating with a dipole layer extending between the immobile oxygen and zinc ions.

Zinc oxide absorbs strongly in the ultraviolet for an energy of 3·2 eV (380 mμm) which corresponds to its band gap. The absorption coefficient at this energy for an Electrofax layer although considerably less than that of zinc oxide still has the high value of 7×10^3 cm^{-1}. Thus illumination of an Electrofax layer with band gap radiation generates electron–hole pairs down to a depth of 4 μm and the majority of these pairs will be located near the surface of the layer. The positive holes migrate to the surface of each crystallite and neutralize the oxygen ions, the resultant neutral atoms or molecules desorb from the surface, although the resin tends to inhibit movement too far from a given crystallite. The electrons that are produced by illumination have a very long lifetime compared with the holes (10^{-5} sec), however they rapidly fall into shallow traps and remain in thermal equilibrium with the conduction band. The figure of 1 in 10^9 has been estimated as the fraction of electrons which are produced by exposure to light and that remain in the conduction band. This increases the time for complete recombination of electrons by a factor of 10^9, once the illuminating source is removed, and several hours are usually required for a return to the dark adapted state. In the dark, the photodesorbed oxygen diffuses back to the surface and the highly insulating state is reached when resistivities of 10^{16} Ω-cm or more are reached. Exposure to infrared radiation hastens the dark adaption by the rapid emptying of occupied donor levels.

The corona charging is a technique used to prepare Electrofax prints and in the technique a large negative D.C. potential (6000 V) is applied to the front face of the layer with the back layer grounded. Negative oxygen ions are then accelerated under this potential and land on the front surface of a dark adapted layer. Negative charge thus accumulates on the front surface, electrons in turn are forced away from the grounded paper support and a positive charge builds up in the support to balance the negative charge deposited. Thus there is seen to be a marked similarity between the effects of corona discharge and dark-adaptation; the principal difference being that in dark adaptation the oxygen ions are distributed uniformly on the crystallite surface, whereas in corona discharge the oxygen ions are deposited on the front surface of the Electrofax layer with positive charges on the back surface. The potential across the layer which results from the corona drops rapidly for a short time and then stabilizes at what is termed the saturation potential. Dark discharge of Electrofax layers which have been corona charged takes place over periods of time ranging from minutes to days which depend on the temperature and the humidity rather than the content of the layer. Photodischarge on the other hand is a more rapid process and even in the presence of a high density of traps is of the order of seconds. The large difference in dark and light discharge times is accounted for in terms of the electrons bound to the surface layer oxygen ions, which form good blocking contacts; the photogeneration creates the mobile carriers to discharge the blocking layer.

7.3.3. *Image Formation in Electrofax Layers*

The latent images formed in Electrofax layers can be of two types, either electrostatic or conductivity, depending on their method of formation. Uniform corona charging followed by exposure to an illuminated object creates a latent electrostatic image on the layer. The saturation potential exists in the unilluminated areas while the potential in the exposed areas depends on the light intensity and hence the number of photogenerated carriers. Thus an electrostatic charge pattern gives a latent image which relaxes in the dark at normal temperatures over a period of hours. The method of development employed for electrostatic images is usually of the dry type with a triboelectric powder and metal filings.

Latent conductivity images are formed by exposure of uncharged, dark adapted layers to an illuminated object. Most of the photoexcited electrons move into local trapping states. Application of an electric field extracts the few electrons present in the conduction band. These are replaced by thermally excited electrons from shallow traps which are in turn extracted. The electrons which thus form the latent image are just those originally excited by the photoabsorption. Development can be effected by both dry and wet means; charged metal ions under the influence of an applied electric field are deposited on the surface to give darkening.

Refined techniques are used to produce Electrofax prints which involve the formation of both latent conductivity and electrostatic images before development. However, even in desensitized Electrofax layers the resultant prints are more contrasty than a normal contrast silver halide photographic print. The sharp contrast which occurs with Electrofax layers coupled with single stage processing underline the advantages of this method of reproduction.

7.4. Space Charge Limited Devices

7.4.1. *Basic Device Principles*

Space charge limited (SCL) devices which use CdS and CdSe as basic materials have been investigated in considerable depth; references

20–24 give an indication of the degree of development. The operation in dielectrics under SCL conditions suggests short carrier transit times with consequent good high frequency performance, relative insensitivity to temperature and low noise. Further, devices that operate with carriers of one sign impose less stringent requirements on the material's purity.

Both diodes and triodes have been constructed which use CdS coupled with appropriate metallic electrodes. Indium or aluminium provide an ohmic contact and gold is satisfactory in most instances as a blocking

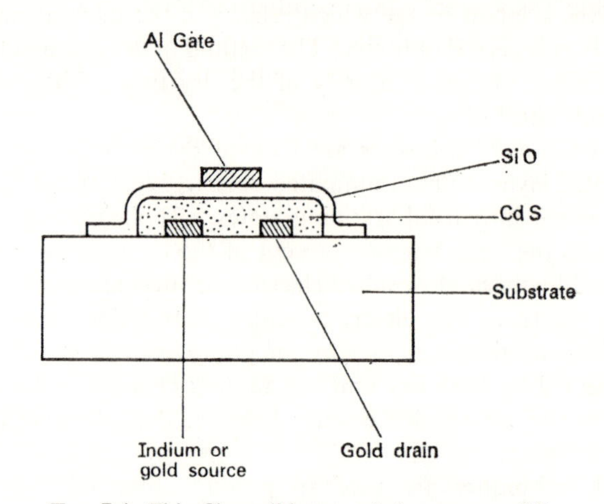

FIG. 7.4. Thin film solid state triode structure.[22]

contact on CdS. Single-crystal platelets and polycrystalline thin films of CdS have been used successfully to fabricate devices. Investigations of the characteristics of SCL diodes seem to indicate that no frequency cut-off results from transit time effects. Hence the frequency limitation is determined only by the accuracy with which the device capacitance can be neutralized and this is likely to permit operation at submillimetre frequencies.

The electrode configuration in the SCL solid triode differs from the vacuum equivalent since the inclusion of electrodes within the thin film or crystal is very difficult to achieve. Reference 25 gives a detailed account of the behaviour of solid state SCL devices and their incorporation in electric circuits. In Fig. 7.4 the cross-section of an insulated gate CdS triode is shown. This is an example of a SCL device in which the insulated gate is used to provide an extremely high input impedance of between 10^{12} and 10^{15} ohms; impedances of this magnitude are not obtainable with the conventional ambipolar transistor. Contrary to the usual practice gold, despite its high work function, gives ohmic contacts with CdS thin films if it is evaporated onto the substrate before the CdS.[(23)] The mechanism for this effect is that during evaporation the CdS is highly dissociated, consequently free cadmium can alloy with the gold layer to form a clean ohmic contact to the CdS film. More recently gold has been found to form an ohmic contact to CdS crystals which have been subjected to strong ion bombardment.[(26)] The CdS film thicknesses are generally of the order of one micron. The substrate temperature is held at 180°C to prevent excess cadmium in the deposited layer and under these conditions thin film resistivities of 10^8 Ω-cm are obtained.

7.4.2. *Integrated Thin Film Scan Generator*

Although the technology of thin films of semiconducting compounds has not reached the advanced stage that it has in bulk elemental and compound semiconductors, nevertheless startling advances have been made with thin film circuitry. Weimer *et al.*[(27)] have produced a 180-stage integrated thin film scan generator, the function of which is to test solid state image sensor panels. This generator was produced in one pump down of the system and used a composite set of masks to produce the pattern. Composite masking overcomes the main objections to single-layer mechanical masking which in the case of complex patterns leads to poor image definition and subsequent unreliability unless extreme precision in manufacture, usually at a prohibitive cost, is achieved. With the composite set up effectively three masks are employed, the one closest to the thin film provides the precise definition in one direction while the other two act as limiters at right angles to each other.

The evaporated circuit produced by Weimer was composed of 360 thin film transistors, 180 field effect diodes, 360 resistors and 180 capacitors and required some 20 to 30 individual evaporations. Cadmium selenide was used as the active component, gold plus indium as source and drain, silicon monoxide for the capacitors and insulated gates in the transistors, aluminium for the gate electrode and nichrome for the resistive components. Thin film CdSe transistors were

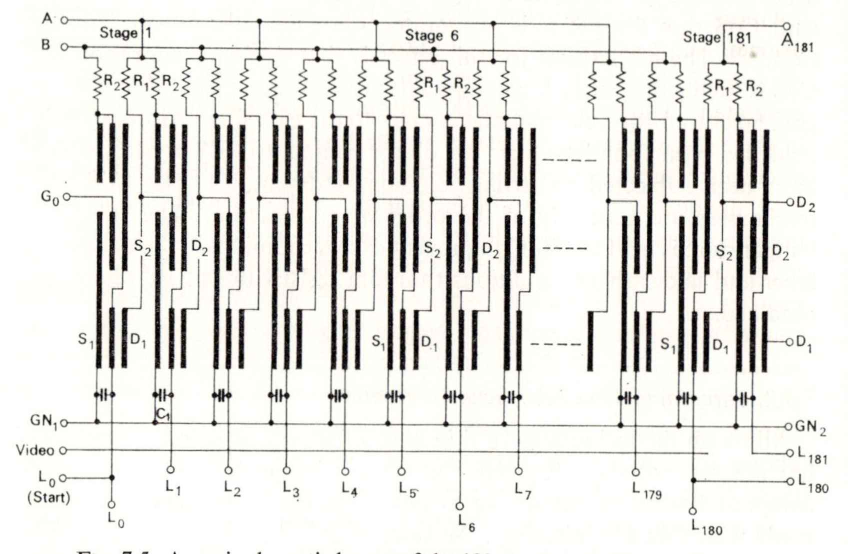

FIG. 7.5. A semi-schematic layout of the 180-stage scan generator showing external connections to the peripheral thin film monitoring transistors.[27]

evaporated round the periphery of the scan generator so as to electrically monitor the set up during fabrication. Figure 7.5 illustrates the scan generator with connections for the monitoring devices. An earlier piece of work by the same investigators on a 30-stage scan generator had taken CdS as the active component; this required a heated substrate because of the preferential loss of sulphur. With CdSe no such heating was necessary. The scan generators that have been produced up to the present time have given output voltage pulses of from 2 to 4 V and operated at clock frequencies ranging from 12 Kc/s to 2 Mc/s without a load. Improvements in the design of circuits should extend the operating frequencies at both ends and perhaps at higher frequencies approach the current frequency test limit in evaporated thin film transistors of 75 Mc/s.

7.5. Gunn Effect Devices

Gunn[28] discovered that when the D.C. bias field across randomly oriented short *n*-type samples of GaAs and InP exceeds a critical value of several thousand volts per cm, coherent microwave output is generated. The frequency of the output, which was initially observed in the 1–5 Gc/s range, is related to the product of the velocity of the electrons computed from the low field mobility and the threshold bias field, divided by the length of the sample. Subsequent development of the effect has produced a device with a carrier wave (CW) output of high spectral purity;[29,30,31] D.C. to microwave conversion efficiencies up to 5% and output powers of 60 mW (CW) and 100 W (pulsed) have been obtained. Negative conductance amplifiers constructed from Gunn effect materials indicate the profitable exploitation of the special physical properties of these materials can have a significant impact on microwave technology. Although the power levels, quoted above, are high for solid state devices, it is likely to be some time before these compete with the high power electron tubes.

Gunn demonstrated that the microwave output is correlated with the transit time across a given sample of a high field domain associated with a moving dipole layer of charge. These domains are a consequence of the voltage controlled bulk negative conduction mechanisms.[32] One of these mechanisms is based on the electric field dependent excitation of carriers from a low-lying high mobility valley to normally unoccupied low mobility valleys. In GaAs it has been shown that transition from the high mobility (000) minimum to the low mobility (100) valleys requires an energy of 0·35 eV.[33] This model has been supported by studies with large hydrostatic pressures applied to lower the energy separation of these minima and the threshold field for onset of the effect.[34]

N-type CdTe has been found to exhibit the Gunn effect with the same general features as in GaAs.[35] The sample current increases linearly with voltage at low fields, but the increase is much slower for fields in excess of 13,000 V/cm. The current oscillates from the threshold to lower values for voltages above threshold. The threshold electron drift velocity is $1{\cdot}3 \times 10^7$ cm/sec, similar to that in GaAs. The threshold field necessary in CdTe for oscillation to occur is approximately four times that in GaAs and is a measure of the energy separation of the two conduction band minima. Hydrostatic pressure measurements to locate the higher energy minimum and associated mobility indicate an impurity level above the lowest minimum is involved. The extension of the frequency range into the X-band region and beyond has already been achieved with thin wafers of GaAs as far as 30 Gc/s. However, the spectral purity in this region is limited since the surface finish is approaching the wafer thickness.[36]

7.6. Ultrasonic Amplification

The development of ultrasonic amplifiers has centred around three of the II–VI compounds, ZnO, CdS and CdSe.[37, 38, 39] Amplification of ultrasonic waves occurs when the drift velocity of optically excited electrons exceeds the velocity of sound in a crystal. The experimental arrangement of the amplifier is shown in Fig. 7.6. The transducer converts a short duration pulse (1 μs) of R.F. signal (15–400 Mc/s) into ultrasonic waves. These waves travel through the active crystal under the influence of a D.C. or pulsed drift field to the second transducer which in turn converts the sound wave back into an R.F. signal. In these crystals strong piezoelectric coupling exists between the mechanical stress and strain and the electric field and displacement, when the sound and electronic carrier velocities are similar. When the electron velocity exceeds the sound velocity absorption of energy from the electrons occurs and amplification of the sound wave is observed. Departure from ohmic behaviour in the piezoelectric semiconductor has been observed at the threshold field for ultrasonic amplification. This has been shown experimentally to be due to ultrasonic flux build up and the accompanying acousto–electric current. In order to reduce losses in the system, the end surfaces of the crystal are highly polished and indium layers are used to bond the surfaces to the quartz transducers. To reduce losses still further an integrated amplifier has been produced.[40] In this conductive and transducer layers are formed on the surfaces of a piezoelectric CdS crystal by diffusion techniques. Heat treatment in a cadmium rich atmosphere creates a conductive layer, then diffusion of copper causes compensation of donor levels and subsequently a high impedance transducer layer.

Both shear and longitudinal wave amplification has been obtained in these crystals; however, the shear wave mode is the simplest to

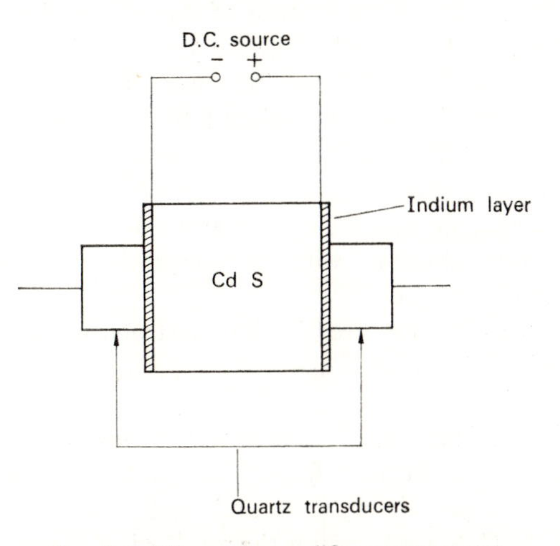

FIG. 7.6. Ultrasonic amplifier arrangement.

produce since lower drift fields are required and consequently less power is dissipated in the amplifying mode. In CdS the threshold drift field for amplification in the shear wave mode is 700 V/cm and this corresponds to an electron drift velocity equal to the shear wave velocity of 2×10^5 cm/sec. It is necessary to illuminate the CdS crystals to photoexcite the requisite number of carriers for sufficient gain to be observed. The resistivity of the CdS crystals under operational conditions is of the order of 10^4 Ω-cm which gives a power dissipation of 100 W. Because of these high power levels, pulsed operation is preferred in such amplifiers. The physical size over which this power is dissipated is generally less than a cubic centimetre. The actual gains obtained can be as large as 100 dB/cm at 60 Mc/s, although losses, which occur in both the transducers and in the coupling, lead to a reduction in the effective gain. A gain of 35 decibels at 54 Mc/s has been obtained with the integrated amplifier.[40]

7.7. Solar Cells

Solar cells represent an application of the photovoltaic effect which has been investigated for well over ten years. Silicon and cadmium sulphide were the first materials on the scene and in the course of time have been in competition with cadmium telluride, gallium arsenide and gallium phosphide. Silicon and gallium arsenide solar cells are straightforward *p–n* junction photovoltaic devices in which absorption increases the minority carrier density. The formation of *p–n* homojunctions in CdS has not been observed with any certainty and consequently a surface cell must be formed.

Strongly *n*-type CdS was used in the construction of the early solar cells which had an evaporated layer of copper on the back face of the CdS crystal. Indium provided an ohmic contact to the front crystal face. Radiation incident on the cell travels through the CdS layer and generates electrons at the rear face. The photoresponse of this type of cell lies in the range of 1 to 2·4 eV and suggests that photoionization of electrons at the CdS–Cu barrier is the mechanism of carrier excitation. The minimum energy of 1 eV can be ascribed to the barrier height and the high energy cut off at 2·4 eV to absorption in the CdS. These CdS cells show a similar dependence in performance with temperature to conventional *p–n* junction cells but have a much lower conversion efficiency; thus they do not offer the anticipated high temperature advantage over silicon and germanium.

More recent investigation of cadmium sulphide[41] and cadmium telluride solar cells[42] have been in the direction of a cell which depends more on the bulk properties of the semiconductor rather than metallic barrier effects. The front face of the cell is of more importance in this form of the device. Both single crystals and thin films have been used and it is here that basically CdS and CdTe may have the edge on silicon and gallium arsenide. The cost of fabrication of a *p–n* junction in a single crystal form is much higher than in a thin film multilayer device even if the latter suffers from much poorer efficiency. The large area obtainable with thin films can more than compensate for the loss of efficiency. Both *n*-type CdS and CdTe films require a *p*-type junction layer at the front surface. The *p*-type layer is produced by the reaction

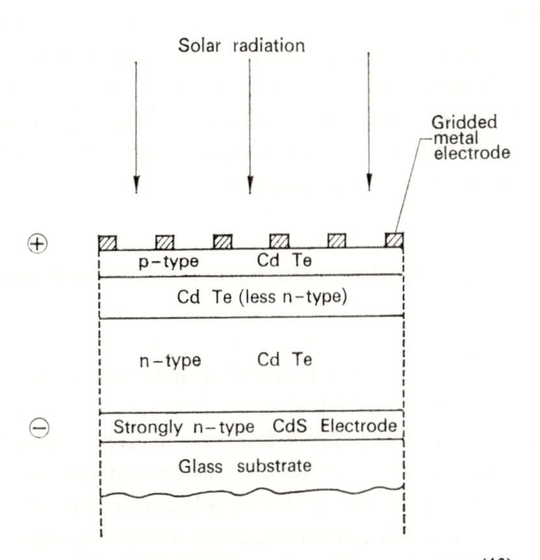

FIG. 7.7. CdTe thin film solar cell arrangement.[42]

of the front electrode which is of copper with the surface layer of the semiconductor; in the reaction either Cu_2S or Cu_2Te is formed depending on whether the sulphide or telluride is the bulk *n*-type layer. The Cu_2S or Cu_2Te is *p*-type and thus a *p–n* heterojunction can exist. A deposited layer of Cu_2O has been found equally effective as a *p*-type layer on cadmium sulphide.[41]

Cusano[42] treats the *p–n* heterojunction in some detail and concludes the solar conversion efficiency of CdTe solar cells can easily be raised to 10%. In the CdTe cell the Cu_2Te is thought to be an indirect gap semiconductor which consequently has a low absorption for incident radiation. Hence most of the absorption occurs in the CdTe and does not accordingly seriously limit the conversion efficiency. It is of interest that the CdTe thin film cell, illustrated in Fig. 7.7, has a strongly n-type layer of CdS ($\rho \sim 0{\cdot}01$ Ω-cm) to provide its back surface ohmic contact. Solar conversion efficiencies of up to 6% have been obtained with this arrangement for thin films and $7\frac{1}{2}$% for single crystals. Cadmium telluride has an advantage as a solar energy converter over cadmium sulphide in that its peak spectral response is very much closer to the peak in the solar spectrum. The absorption of solar radiation of lower energies can be effected by the inclusion of suitable impurities which permit two step optical processes in addition to valence to conduction band transitions.

7.8. Thermoelectric, Thermomagnetic and Galvanomagnetic Applications

7.8.1. *Thermoelectric*

The tellurides and mercury selenide have received considerable attention from the point of view of their thermoelectric and thermomagnetic properties. Cadmium telluride which has been prepared in both p- and n-type form has shown the most promising results although the best value obtained for its thermoelectric figure of merit (10^{-3} per °K at 300°K) is a little disappointing. Such a value is a factor of three lower than that of the best known thermoelectric materials, bismuth telluride and bismuth selenide. The maximum cooling powers obtained with CdTe in a Peltier cooler such as illustrated in Fig. 7.8 (a), have only been 2 to 3 W. Goldsmid[(43)] has pointed out that HgTe and HgSe with electron mobilities of approximately 20,000 cm^2/V-sec at 300°K and lattice thermal conductivities of 0·03 and 0·01 W/cm/degree respectively, might be useful thermoelectric materials when alloyed to other tellurides and selenides to give a positive energy gap. In this respect the work of Wright[(44)] on $Hg_5In_2Te_8$ suggests that this might be a suitable material, since it has a mobility of 20,000 cm^2/V-sec at 77°K and a thermal conductivity of 0·01 W/cm-°K.

7.8.2. *Thermomagnetic*

There exists also the possibility that HgTe, HgSe and associated alloys might provide an Ettinghausen cooler of the type shown in Fig. 7.8. This uses the fact that a magnetic field can enhance the figure of merit in a semi-metal or an intrinsic small energy gap semi-conductor. In the Ettinghausen effect electrons and holes are trans-

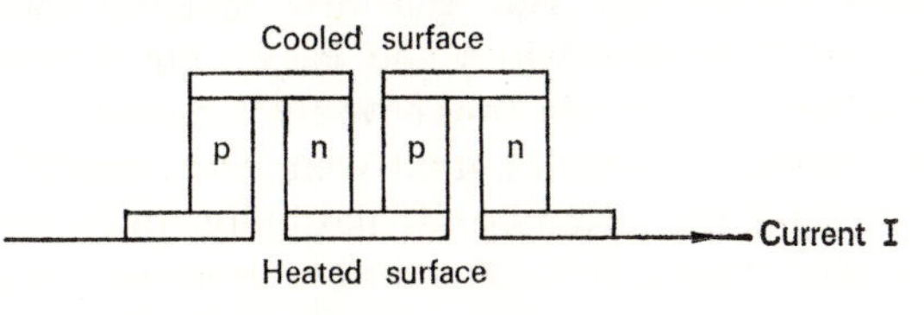

(a) Thermoelectric cooler

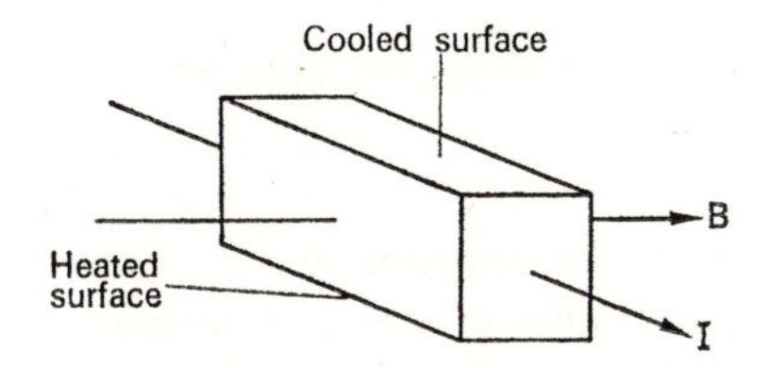

(b) Simple thermomagnetic cooler

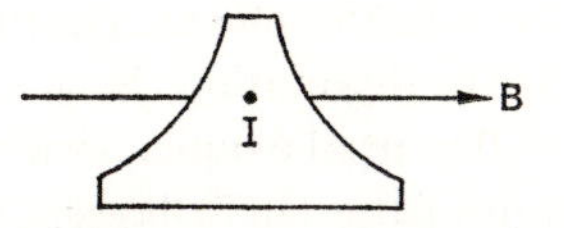

(c) Optimized cross section for thermomagnetic cooler

FIG. 7.8. Thermoelectric and thermomagnetic cooling arrangements.

ported from the face of the sample where they are created by thermal generation to the opposite face where they are annihilated with thermal emission. Energy is thus transported from one side of the sample to the other to give cold and hot faces. Investigations on Bi–Sb alloys have given temperature differences as great as 120°K for a room temperature hot face and this corresponds to a figure of merit of 5×10^{-3}

per °K. HgTe and HgSe both unfortunately have electron–hole mobility ratios of 50 or greater at all temperatures and consequently the creation and annihilation process is very inefficient and results in a reduced figure of merit.

7.8.3. *Galvanomagnetic*

InSb is well known as a Hall generator material and if thin film material is used high sensitivities and large output voltages can be obtained. HgTe and HgSe thin films have similar applications, although they do not provide such high output voltages as InSb.[45] However, these films do have the advantage of flexibility which permits a Hall generator of total thickness 20–30 μm to be bent up to a curvature of 1 cm. The thickness of the active medium is 1·3–1·5 μm which is evaporated onto a mica substrate. With these mechanical properties it is possible to measure field strengths in extremely narrow and curved gaps. Recent investigations into HgTe thin films indicate that the elastic properties which allow curvatures of 1 cm are confined to film thickness less than 3 μm., Further, the resistivity of a thin film less than 1·3 μm in thickness increases rapidly with decreasing thickness; hence the optimum thickness range for these generators is taken to be 1·3–1·5 μm to combine the best electrical and mechanical properties. The order of magnitude of sensitivity obtained is 50 μV/Weber/cm^2 for an input current of 1000 A/cm^2. The area of the thin film is approximately $0{\cdot}3 \times 0{\cdot}6$ cm^2. Heat dissipation is a considerable problem with the active film since differential expansion with respect to the mica substrate can cause rupture of the HgTe layer. The value of 130°C is quoted as the maximum temperature increase which is permissible over the temperature at which the layer is evaporated; this can be shown to set an upper limit to the input current for a given geometry.

7.9. Image Intensifiers

The combination of electric field and photon sensitive properties has permitted the application of CdS, CdSe, ZnS and ZnSe in several fields, which include image intensification, image processing, logic networks and shift registers.[46] A particular example of these electro-optic devices where much progress has been made is the image intensifier. Single and double layer image intensifiers have been produced, and it is the latter type on which work is being undertaken at present.

7.9.1. *Single Layer Intensifier*

The single layer intensifier uses a phosphor layer sandwiched between two electrodes, one of which is transparent. A D.C. electric field of about 10^5 V/cm is applied to the phosphor layer and an image

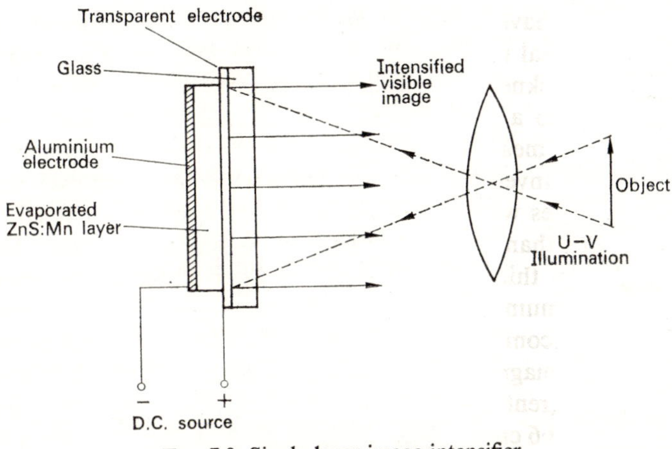

FIG. 7.9. Single layer image intensifier.

focused on to the surface of the layer. The very small electroluminescence emission of the phosphor layer is enhanced by the incident light and a light gain of 10 has been obtained by Cusano.[47] In the case of ZnS, ultraviolet or X-ray illumination of the object is necessary in order to excite the phosphor layer. Kazan and Nicoll[48] have used an evaporated layer of manganese activated zinc sulphide, 10 microns thick, and a TiO_2 conducting glass as the transparent electrode to observe single layer image intensification, Fig. 7.9. In this device the input–output ratios dependence on intensity is always greater than unity, thus the contrast in the intensified image decreases with intensity. The response times of the ZnS layers are unfortunately too long to observe moving images. Cusano has found that although rise times are very rarely less than 1 s, decay times can be extended to several minutes by the choice of materials and thus permit image storage. The images produced have high resolutions as much as 2500 elements per inch and provide good quality pictures.

7.9.2. *Double Layer Intensifier*

The double layer intensifier is composed of a photoconductive and an electroluminescent layer in series with an A.C. potential applied across the combination. The photoconductor acts as the current control in the circuit and regulates the electroluminescent emission. In practice a narrow opaque layer separates the photoconductive layer from the electroluminescent layer in order to prevent optical feedback from the luminescent emitter. The simple theory of the device shows that the photoconductive layer must be many times thicker than the electroluminescent layer for a large gain to occur. The region of high gain for this system occurs only over a small range of intensity at a given frequency. It is possible to operate the system with a combination of two frequencies and increase the intensity range over which gain occurs; also some linearization of the gain is obtained under these conditions. The simplest form of the double layer device is the sandwich-type similar to Fig. 7.9, except that the exciting radiation falls on the photoconductor. Panels 12 in. square have been constructed and these have found particular application in X-ray fluoroscopy, where the image intensity of the standard fluorescent screen has been increased one hundred fold and permits the X-ray dosage to be lowered.

There exist practical limitations with these sandwich type intensifiers in that visible radiation is not transmitted through the CdS layer but merely produces surface currents. The photoconductive layer if made in a grooved form overcomes this difficulty, although the requirement of numerous electrodes reduces the simplicity of the device somewhat, Fig. 7.10. Refinements of the grooved intensifier have lead to increases in the gain of 50 from the arrangement of Fig. 7.10. Alternate electrodes have negative or positive D.C. bias, so that the A.C. applied voltage does not change sign on either set of electrodes. Under these conditions the best photocurrent is obtained and the phosphor remains under A.C. excitation. Image erasure is effected by reversing the current and the normal decay time is shortened greatly.

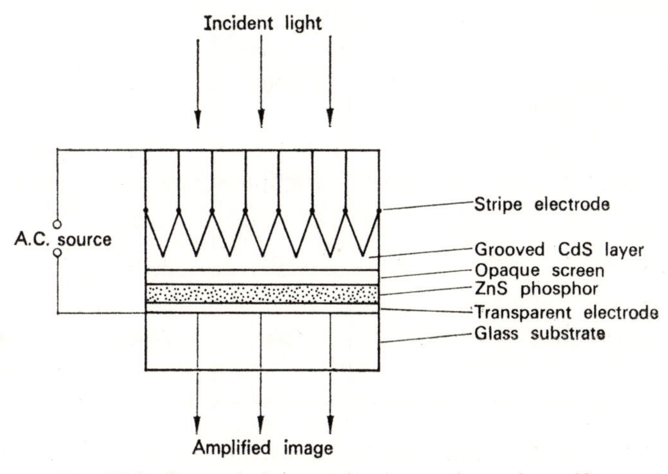

FIG. 7.10. Grooved photoconductor type image intensifier.

7.9.3. *Complex Layer Intensifier*

Both two colour input and output image intensifiers have been constructed. These can use cadmium sulphide and selenide as the photoconductive materials and zinc sulphide and zinc selenide as the phosphors. The spectral responses of the photoconductors differ considerably and a combination of the two responses permits the spectrum from 500 to 900 mμm to be completely covered. The method of construction makes alternate grooves of CdS and CdSe and the associated phosphor layers are separated by the opaque insulation layer as shown in Fig. 7.11.[49] The gain of this device fluctuates between 10 and 200 depending on which photoconductor is activated and the intensity range that is used. CdS has the greatest gain but this falls rapidly at both low and high intensities. CdSe, although it has a

TABLE 7.1. CdSe SINTERED LAYERS OPERATION IN AN IMAGE INTENSIFIER[46]

Operating frequency (c/s)	10,000	4200	2000	420
Input-light level (ft-candles)	0·4–0·02	0·2–0·007	0·1–0·003	0·01–0·0008
Output at maximum input (ft-lamberts)	160	150	140	35
Rise time to 80% (sec)	0·06–0·5	0·07–1·0	0·07–2·0	0·4–6·0
Decay time to 20% (sec)	0·02	0·05	0·09	0·28
Maximum tungsten light gain	555	1900	3400	6000

smaller overall gain, this is almost constant over several decades. In practice the best performance in image intensification in terms of speed of response and high gain has been obtained with sintered CdSe. Nicoll[50] has compared the operating range of CdSe sintered layers

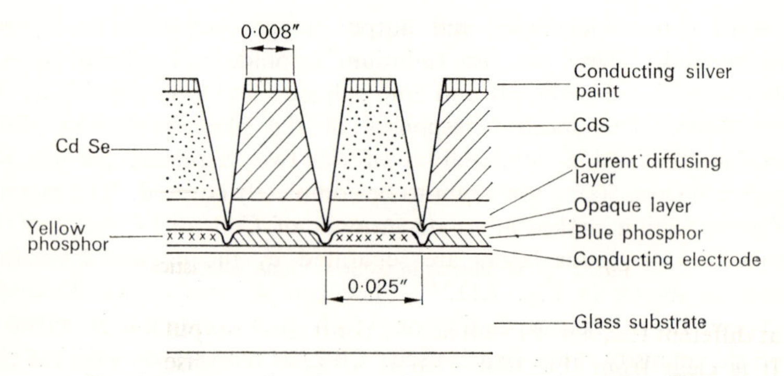

FIG. 7.11. Two-colour input and output image intensifiers.[49]

TABLE 7.2

Panel type	Photoconductor material	Threshold input ft-candles	Exposure for max. gain ft-candles-sec	Gain	Recovery time between images, sec
Amplifier	CdS (Powder)	0·003	0·3	800	15–60
Amplifier	CdSe (Sintered)	0·001	0·01	1200	1
Storage with feedback	CdS (Powder)	0·02	1·0	200	5
Storage with feedback	CdS (Sintered)	0·02	1·0	200	15
Storage	CdSe (Powder)	0·0005	0·01	2000	0·1

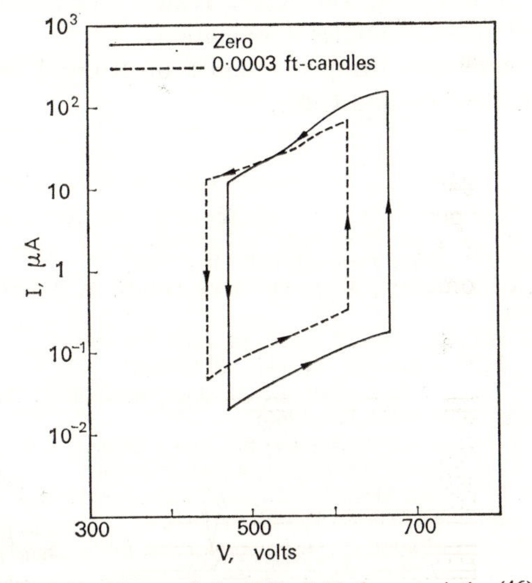

FIG. 7.12. Hysteresis in CdSe I–V characteristics.[46]

at different frequencies with a 50–1 contrast in output image, Table 7.1. It is clear from this table that if a rapid response is required high frequencies must be used, although a considerable loss in gain ensues.

7.9.4. *Image Storage*

Image storage devices have been constructed which in the simplest form are basically the same as the image intensifier except that optical feedback is no longer prevented and is used to give the image storage. The level of feedback which provides periods of storage greater than minutes is of the order of 1 %. In addition to the use of optical feedback as a storage mechanism observations on CdSe layers have indicated that the I–V characteristic has a rapid increase at a high voltage which depends on the illumination level of the photoconductors. This property has been used to sustain a high current and hence an image on the electroluminescent screen. Erasure is achieved by a considerable reduction in the applied voltage which returns the CdSe to a normal I–V characteristic. Such an image storage device thus relies on a hysteresis effect in the photoconductor. Figure 7.12 illustrates this effect for zero and 0·0003 ft-candles illumination.

Table 7.2 compares the performance of grooved image intensifier panels which use CdS and CdSe.[46]

References

1. Kruse, P. W., *Applied Optics* **4,** 687–92 (1965).
2. Nathan, M. I., *Applied Optics* **5,** 1514–28 (1966).
3. Basov, N. G., Grasyuk, A. Z., and Katulin, V. A., *Sov. Phys. Doklady* **10,** 343–4 (1965).
4. Holonyak, N., *et al., Proc. I.E.E.E.* **54,** 1068–9 (1966).
5. Phelan, R. J., *Proc. I.E.E.E.* **54,** 1119–20 (1966).
6. Koniukhov, V. K., Kulevskii, L. A., and Prokhorov, A. M., *I.E.E.E. J. Quantum Electronics* **OE2,** lxv (1966).
7. Hurwitz, C. E., *Appl. Phys. Letters* **8,** 121–4 (1966).
8. Hurwitz, C. E., *Appl. Phys. Letters,* **8,** 243–5 (1966).
9. Morehead, F. F., and Mandel, G., *Appl. Phys. Letters* **5,** 53–54 (1964).
10. Ayen, M., *Appl. Phys. Letters* **7,** 146–8 (1965).
11. Mandel, G., and Morehead, F. F., *Appl. Phys. Letters* **5,** 143–5 (1964).
12. Morehead, F. F., *J. Appl. Phys.* **37,** 3487–92 (1966).
13. Weiser, K., and Woods, J. F., *Appl. Phys. Letters* **7,** 225 (1965).
14. Crowder, B. L., Morehead, F. F., and Wagner, P. R., *Appl. Phys. Letters* **8,** 148–9 (1966).
15. Schiel, E. J., Bullwinkel, E. C., and Weiner, R. B., *Proc. I.E.E.E.* **53,** 2140–1 (1965).
16. Kelly, C. E., *Trans. I.E.E.E.* **ED12,** 1–4 (1965).

17. Hamilton, G. W., and Fowler, A. L., *Electronics and Power* **12,** 318–22 (1966).
18. Young, C. J., and Greig, H. G., *R.C.A. Rev.* **15,** 469–84 (1954).
19. Amick, J. A., *Photoelectronic Materials and Devices*, pp. 373–424. Van Nostrand, New Jersey, 1965.
20. Wright, G. T., *Proc. I.E.E.E.* **51,** 1642–52 (1963).
21. Shallcross, F. V., *Proc. I.E.E.E.* **5,** 851 (1963).
22. Zuleeg, R., *Solid State Electronics* **6,** 645–55 (1963).
23. Zuleeg, R., and Muller, R. S., *Solid State Electronics* **7,** 575–82 (1964).
24. Page, D. J., *Solid State Electronics* **9,** 255–64 (1966).
25. Sevin, L. J., *Field Effect Transistors.* McGraw Hill, New York, 1965.
26. Robertson, J. M., and Ray, B., *Physica* **33,** 108–12 (1967).
27. Weimer, P. K., Sadasiv, G., Meray-Horvath, L., and Homa, W. S., *Proc. I.E.E.E.* **54,** 354–60 (1966).
28. Gunn, J. B., *Solid State Comm.* **1,** 88–91 (1963).
29. Gunn, J. B., *I.B.M. J. Res. Dev.* **8,** 141–59 (1964).
30. Haaki, B. W., and Knight, S., *Solid State Comm.* **3,** 89–91 (1965).
31. Quist, T. M., and Foyt, A. G., *Proc. I.E.E.E.*, **53,** 303 (1965).
32. Ridley, B. K., *Proc. Phys. Soc.* **82,** 954–66 (1963).
33. Hilsum, C., *Proc. I.R.E.* **50,** 185–9 (1962).
34. Hutson, A. R., *et al., Phys. Rev. Letters* **14,** 639–41 (1965).
35. Foyt, A. G., and McWhorter, A. L., *Trans. I.E.E.E.* **ED13,** 79–87 (1966).
36. Mao, S., and Pucel, R. A., *Proc. I.E.E.E.* **54,** 414–15 (1966).
37. Hutson, A. R., McFee, J. H., and White, D. L., *Phys. Rev. Letters* **7,** 237–9 (1961).
38. McFee, J. H., *J. Appl. Phys.* **34,** 1548–53 (1963).
39. Midford, T. A., *J. Appl. Phys.* **35,** 3423–5 (1964).
40. Chubachi, N., Wada, M., and Kikuchi, Y., *Jap. J. Appl. Phys.* **3,** 777–9 (1964).
41. Grimmeis, H. G., and Memming, R., *J. Appl. Phys.* **33,** 2217–22, 3596–7 (1962).
42. Cusano, D. A., *Solid State Electronics* **6,** 217–32 (1963).
43. Goldsmid, H. J., *Thermoelectric Refrigeration.* Heywood, London, 1964.
44. Wright, D. A., *Brit. J. Appl. Phys.* **16,** 939–42 (1965).
45. Kobus, A., and Ignatowicz, S., *Solid State Electronics* **9,** 595–600 (1966).
46. Nicoll, F. H., *Photoelectronic Materials and Devices*, pp. 313–73. Van Nostrand, New Jersey, 1965.
47. Cusano, D. A., *Phys. Rev.* **98,** 546–7 (1955).
48. Kazan, B., and Nicoll, F. H., *J. Opt. Soc. Am.* **47,** 887–94 (1957).
49. Nicoll, F. H., and Sussman, A., *Proc. I.R.E.* **48,** 1842–6 (1960).
50. Nicoll, F. H., *R.C.A. Review* **20,** 658–69 (1959).

INDEX